Internal Factors in Evolution

Internal Factors in Evolution
Lancelot Law <u>Whyte</u> ✓

George Braziller, New York

Contents

Glossary

Coordinative Conditions (C.C.)	The mathematical expression of the general conditions of biological coordination; the rules of geometrical and kinematic ordering which must be satisfied (to within a threshold) by the internal parts and processes of any viable cellular organism.
Genotype	The complete system of factors, in genes, chromosomes, cytoplasm, and cell-cortex, making up the genetic constitution of an organism.
Phenotype	The system of manifest characters of an organism.
Mutations	Discrete inherited changes in the genotype.
Internal (*or developmental*) Selection	The internal selection of mutants at the molecular, chromosomal, and cellular levels, in accordance with their

7

Glossary

External
(*or Darwinian
adaptive*)
Selection

compatibility with the internal coordination of the organism. The restriction of the hypothetically possible directions of evolutionary change by internal organizational factors.

A net reproductive differential of genetic variants caused by differences of adaptive fitness of competitive individuals. The further restriction by external factors of the avenues of evolutionary change permitted by internal selection.

8

Argument

Preface: A biological surprise—in addition to Darwinian adaptive selection another mechanism of selection has been a directive factor in the evolution of species: a direct selection of mutants in accordance with the system's capacity for coordinated activity—*Darwin structured* and thereby transformed—why I enter this difficult field—my interest in ordered structures—two physical concepts: *forces* and *order*—studies in biology—my view of internal factors expressed in 1949—stimulus from J. B. S. Haldane in 1959—idea taken up again—papers published in 1960 and 1964—many simultaneous statements of this collective idea—neglect in text books— why write a book?—a case history in the development of scientific ideas.

1: Biological Organization. Three basic ignorances in science—the problem of biological organization central —this defined—the views of Aristotle, Descartes, and the

Argument

new organismic school—intense research but inadequate theory—what is missing: a mathematical formulation of the problem—the Coordinative Conditions
(C.C.) defined—their properties and scope—problems
which cannot be solved until the C.C. have been identified.

2: The Synthetic Theory of Evolution. Matured 1930–50
—a magnificent collective achievement—a necessary but
not a sufficient theory of the directive factors—the production of variation and the choice of variants—the
theory analyzed—the skeleton of the argument—five
criticisms considered—the theory undoubtedly contains
one important aspect of the truth and serves as a basis
for the next advance.

3: Internal Factors. Importance of ultra structure—no
arbitrary changes in the genetic system physically, chemically, or functionally stable—stringent internal restrictions on permissible mutations—the mutated genotype
and its consequences must conform to the C.C.—the
discrete spectrum of permissible variations—mutation
"selection rules"—the nature of life limits its variation
—internal selection a second directive factor in phylogeny—elimination of non-conforming mutants—three
possibilities—internal and external selection essentially
distinct—the insides of organisms unlike the spaces between them—the difficulty of fully understanding a new
idea in all its implications—purpose of this essay.

4: Internal Factors (Cont.) Importance of a structural approach to ontogeny—three common errors—Darwin's assumption that "many more individuals are born than can possibly survive" may not be valid when an entirely new type emerges—internal selection may be of importance in macroevolution—the combination of internal and external selection far from obvious—this illustrated in a diagram—ignorance of details does not always matter—Einstein on strong formal hypotheses—in special circumstances internal factors may not merely select from given variations but directly determine successful variations.

5: History of the Idea. Value of historical surveys—two periods: 1880–1900, and 1940 onwards—Roux and Weismann—their ideas not a true anticipation—Huxley in 1888 and Morgan in 1919—the preparation of the ground during 1930–1940 for the study of internal factors—vague early statements—four positive formulations, 1949–1952, by Spurway, Whyte, Bertalanffy, and Lima-de-Faria—subsequent statements from varied points of view—Waddington and Darlington—recent suggestions.

6: Difficulties Answered. The paradox: an idea which is a commonplace truism for some is neglected by others and absent from the textbooks—a spectrum of points of

Argument

view expressed privately 1959–1963—reason for neglect in published works: *no leader yet recognizes the importance of internal selection as a novel principle potentially affecting the entire interpretation of evolution* —some persistent difficulties reconsidered—internal selection not a mere extension of Darwin's mechanism to the internal environment—the idea not premature— many lines of research already require it for their interpretation—the idea not damaging to the precision or objectivity of biology.

7: Conclusions. Twelve provisional but well-supported conclusions put forward for the consideration of specialists.

8: Questions for Tomorrow. Research on biological organization and on the C.C. will take place in the context of a developing physics, biophysics, and biochemistry—towards a science of complex partly ordered changing systems—the C.C. must cover *all aspects of the unity of organism*—this the key problem—six questions put to specialists—the analogy of crystals and organisms exploited to throw light on the character and role of the C.C.—the principle of internal selection is timely as the expression of the contemporary experimental and theoretical concern with structure.

Preface

"Expect surprises!" should be the watchword of all scientists who try to look beyond the fashions of the day. For every rule has its limits and every concept its ambiguities. Most of all is this true in the science of life where nothing quite corresponds to our ideas, similar ends are reached by varied means, and no causes are simple. Exact biology is advancing apace, but it has not yet exhausted the secrets of life in any field, and no generalization is safe from exceptions. In our day biological dogmatism is silly. But how easy it is to know this in the abstract and how hard it is to hold the mind open to surprises!

This book is about a surprise: a biological idea which is in the air, for it has come to many minds during the last fifteen years, but is still so new and strange and so little understood that few have recognized its far-reaching importance. It may prove to be one of the most fertile scientific ideas of this century, yet it is still un-

13

Preface

noticed except by a few who are nursing, almost unawares, the first stages of a profound revolution in biological thought.

While this idea is a truism to some, it is a discovery of the first importance for anyone who realizes its far-reaching implications. Thus a fertile new idea of the utmost importance is being regarded as trivial by those who do not see all that it involves.

The idea is that in addition to Darwinian selection another selective process has also played an important role in determining the evolution of species. It is now being suggested that beside the well-established external competitive selection of the "synthetic" theory of evolution, an internal selection process acts directly on mutations, mainly at the molecular, chromosomal, and cellular levels, in terms not of struggle and competition, but of the system's capacity for coordinated activity. The Darwinian criterion of fitness for external competition has to be supplemented by another: that of good internal coordination. Internal *co*adaptation is necessary as well as external adaptation. This new idea is *Darwin structured,* and thereby transformed into something different. For a new directive agency has come into sight.

The purpose of this book is to discuss this idea from a general point of view, to describe its history as a collective product of many minds, and to consider some of its implications. No attempt is made to cover the many special aspects which require consideration by biochemists, geneticists, embryologists, and others.

But why does a person mainly interested in physics, the philosophy of science, and the history of ideas, dare to approach the complexities of evolution? Is it not better to leave evolutionary theory to those who have devoted their lives to the study of organisms? The reader will forgive a brief personal explanation which bears on the recent history of scientific ideas.

Much of the scientific thought of the xixth century was influenced by the conception of *kinetic atomism,* of material units moving at random. The corresponding conception representative of the xxth century is that of *ordered structure,* of relatively stable arrangements of units. This far-reaching contrast has not yet reached its climax. For this century has not completed its task of identifying ordered structures wherever they exist and of discovering their laws.

Crisp summaries are dangerous, for they oversimplify. But this simple contrast between the two centuries can help us to understand our own time, always a difficult matter. For example, it makes it easier to appreciate the difference between the theme of this book and that of *The Origin of Species.* Though they may have little else in common, each book is true to its own period. A hundred years have passed, and the emphasis in evolutionary thought is shifting from the external struggle between organisms to the fine structural coordination inside them. But we must examine this changing outlook in more detail.

On a long view certain trends in physics stand out clearly. From Newton onward *forces,* representing the *external* action of one entity on another, were regarded as fundamental. But with quantum theory around 1926 the emphasis changed to *quantized ordered states* of systems, considered *internally.* Physics, once the study of external dynamical actions, is now becoming more and more the investigation of states of order and disorder within systems. This involves a subtle change of method. Instead of entities acting externally through forces, the internal ordered relations of a system are seen as changing in course of time in accordance with laws that are not yet fully clear in any realm. For the new science of order and disorder is not yet mature.

What are the advantages of this new method? One is that for many purposes we do not need to know the actual forces at work, which are often complex and obscure, provided that we can make an appropriate guess regarding the type of ordering in the system which interests us. In fact *order* is sometimes a more powerful concept than *force,* and it may be the task of the Newton of tomorrow to show that it always is.

As a student I was convinced that the xixth century kinetic theories—mainly of gases—and the philosophies built on them exaggerated the degree of disorder in the universe and tended to underestimate the extent of order. In the early 1920's at Cambridge I imagined that the universe could be regarded as a system of systems, perhaps a hierarchy of partly ordered systems, from the nebulae and the solar system to organisms, crystals, and molecules. There was plenty of order as well as disorder in the universe, and the next task was to identify the order everywhere. The dynamical disorder of ideal gases was merely an extreme limiting case which provided the worst of models for the cosmos as a whole. What was needed was a model of ordered or "organic" mechanism, to use Whitehead's term.

It therefore disturbed me when I found that the leading Cambridge physiologists, for whom I had great respect, were trying to treat the insides of living cells as disordered fluids amenable to such crude classical ideas as ordinary concentration gradients! For I was convinced that cells were elaborately organized structures in which the ions, molecules, and smaller cell parts might vibrate and drift under changing electrical conditions, but seldom diffuse freely as if the cytoplasm were an ideal fluid.

So I set out on the task of reading about organisms and of thinking about them *in terms of ordered structure*. What was the kind of ordering, or of organic coordination, characteristic of all cells, or perhaps of micro-organisms, plants, and animals respectively? No-one knew, and until recently few even understood the implications of this question. I realized that it might be

Preface

necessary to develop a general philosophy of organisms as ordered systems before such questions could be answered. Thus I found myself gradually maturing an attitude to inorganic and organic systems appropriate to the mid xxth century. After twenty-five years this was published as *The Unitary Principle in Physics and Biology* (1949).

The structural approach developed in that book led me to certain conclusions regarding organic evolution, which I will quote (p. 176):

(1) Until a general theory of biological organization has been formulated and confirmed, and the laws governing the number, structure, and arrangement of the hereditary units are known, theories of the origin and evolution of organisms are liable to contain fundamental errors. The description of a past evolutionary process involves statistical and other methodological problems of such an exceptional character as to render advance uncertain until all theoretically prior problems have been clarified.

(2) If the possible changes in the patterns and arrangement of the hereditary units are restricted by symmetry conditions determining their stability, there may be no mutations which can be fully ascribed to "chance." All mutations to new stable patterns may necessarily possess favorable or unfavorable properties in relation to the self-stabilizing organization of the system. Indeed it may prove that what are now called "chance" mutations, because they are caused by arbitrary external factors, are

none the less necessarily favorable to survival during certain evolutionary periods, necessarily unfavorable during others, and neutral in yet others. *The actual course of the historical process of the evolution of species may have been determined in some periods and in some lines by true Darwinian competitive selection of rela*tively *arbitrary mutations, and in others by an internal tendency for stable changes in the chromosomes to possess a stabilizing effect on the organization of the system.*

(Italics added 1964)

(3) But even these alternatives do not exhaust the possibilities. It may be that the actual historical evolution of species could not have occurred without the constant interplay of both factors:

(a) the internal formative tendency to establish stable symmetrical units and coordinated patterns of such units; and:

(b) external, competitive, post-hoc selection. We may ask the question: if the formative, self-facilitating, and developing property of protein processes is operative in guiding the (ontogenetic) development of the individual organism, has it or has it not played a guiding role in the (phylogenetic) evolution of the species? But no answer can be given at this stage.

At that time, in 1949, I knew of no one else who was thinking about evolution along similar lines. But I need not have felt lonely for, as so often happens, during those very years, 1949–52, *three* biologists were formulating much the same idea in various contexts—as we shall see in Chapter 4. However I did not know this and, believing that the idea was premature, I left it alone.

Ten years later, in 1959, I discovered that J. B. S. Haldane had published an almost identical idea. Though he only gave it a few sentences this was enough to convince me that the time had arrived to develop further the structural approach to evolution. In 1960 I contributed two short Notes to *Science* on the subject and then set about studying the history of the idea. I found to my astonishment and delight that I was far from being alone, for between 1949 and 1960 some *seven* biologists had begun tentatively to air similar ideas. A new school of evolutionary thought was gradually emerging, though no one seemed to be fully aware of this or to appreciate the far-reaching, indeed revolutionary, character of the new principle that was emerging. I therefore began to correspond with leading geneticists and evolutionists, wrote a further paper (*Acta Biotheoretica,* 1964), and then, realizing that the matter needed fuller treatment, started preparing this book.

When the future looks back on the present, when the historians of 2000 A.D. examine the scientific ideas of the 1960's, they will be faced by a paradox that will baffle them if they are naive enough to think that the communication of scientific ideas prospered in the age of Communication Theory. It is true that ideas travel

quickly in physics, but in general specialists are slow
to admit fundamentally new ideas, even in the enlight-
ened 1960's. One of the most fertile scientific ideas of
our time has already been clearly formulated, and for
some—molecular biologists, for example—is already a
commonplace, but many leading biologists show no
signs of having noticed it and *no book on evolutionary
theory has yet discussed it as a new mechanism of
selection.* The struggles of the history of ideas are not
over and done with in this age of conferences, grants,
and travels. My experience with physics had misled me
into thinking that everyone pays as much attention to
new ideas as I try to myself.

The fact, surely rather awkward for some, is that
students and readers of textbooks on evolution are still
being told that Darwinian-type selection is *the* cause
of organic evolution, the only directive factor in phylogeny, though in private discussions and in some unpub-
lished lecture courses the more fortunate are already
learning that it would be unwise and certainly unscientific to exclude the possibility, or to neglect the likelihood, of internal factors also playing an important role.
Indeed for many the question now is: How can internal
factors possibly *not* be highly important, since genes,
chromosomes, and cells are highly ordered systems? Does
this not imply a very different kind of selection of
mutated genotypes than the Darwinian process of the
dominant theory?

Preface

Still, why write a book about it? The steady accumulation of structural observations will soon compel the adoption of the new attitude, the more conservative evolutionists will shortly retire from the field, and a new generation of molecular biologists will ensure the proper recognition of the importance of internal factors. Is it not too much to expect evolutionary specialists, who have long stabilized their thinking and have committed themselves in print, to see that molecular biology has something important to contribute to their subject?

Well, I cannot stand by and see a fertile and timely idea unduly neglected. To me as a xxth century philosopher of science, concerned with the methods of science today and tomorrow, i.e. with ordered structure, the operation of a structural selection of mutated types, quite distinct from Darwinian selection, is a necessary consequence of everything that I believe I know about inorganic and organic nature. I would stake much more than my reputation on that. The logic of nature, as I understand it, tells me that here is a clue of immense importance, whose fertility for biology—allowing for the present rate of advance of biochemical genetics—should be manifest to all within a generation.

This work is an appeal to those who share this view to come into the open and to say so in print, for the better of their science. Alas, science has no Ombudsman, or I would refer to him a dozen recent texts by authors of the highest reputation which have made no refer-

22

ence to this new principle, so that he might admonish the authors and publicly warn their readers. Many words and much money are being spent on the organization of scientific research. Those who are serious about this and believe that deliberate organization and finance are beneficial should see that this new principle gets its share. Posterity, reading this book, will check on them. There should be an international conference on internal factors in evolution.

One further personal word. The synthetic theory of evolution has always seemed to me to be one of the most impressive achievements of the human intellect, a collective scientific product of indubitable validity, if correctly formulated. It can scarcely be doubted by any objective mind that Darwinian external selection has been an immensely important factor in determining the evolution of the forms of life. But multiple causation has been at work here; other directive factors have also played a part; certainly we are still in many respects too ignorant for incautious formulations to be allowed to pass unchallenged. The exaggerations of the over-orthodox must be met by the clearest possible critique. That I have attempted here. *The internal structure of organisms has directly influenced the avenues of phylogeny.* This assertion will be repeated in the pages that follow, but not as often as contrary statements have been made in countless recent works on evolution.

To speak with authority on evolution—not that "authority" is always desirable—one should be up to date on morphology, paleontology, taxonomy, ecology, genetics, cytology, biochemistry, and the biomathematics of coding. Who is? The past history of the forms of organisms is perhaps the most complex and obscure subject that the scientific mind has ever tackled, and it can only be approached by abstractions very distant from the actual events which it seeks to describe. This being so, constructive ideas potentially capable of providing a new and reliable approach to any aspect of the subject must be welcomed, wherever they come from. If they arise not merely from the natural philosophy of structure but also from observations by working geneticists, they are certainly timely and merit careful consideration by all who seek the truth in this treacherous realm.

May I ask the reader to remember that what is most obvious may be most worthy of analysis? Fertile vistas may open out when commonplace facts are examined from a fresh point of view. What is needed to appreciate the force of my argument is not extensive knowledge of facts, or even confidence in logical analysis, but an awareness of the pervasive influence of organized structure in the living realm.

Whatever may be the degree of validity of my argument I offer here a fascinating *case history* in the development of scientific ideas, hot from the frontiers of science where observations provoke conjecture, and conjecture sometimes becomes knowledge. Here this is in the course of happening under our eyes. Much can be learned from this story regarding the nature of science. For what in 1963 is a commonplace truism to some, is plain nonsense to others, and it is hard to say which attitude is more damaging to biology.

I do not understand why no molecular or other biologist has written this book and I have been forced to undertake it.

I wish to thank very warmly the many biologists who over the last five years have allowed me to discuss this theme with them or have kindly answered my letters. As I shall have to criticize certain points of view expressed by them I wish to emphasize how courteously they have treated my enquiries, even when they failed to accept or even to understand my arguments. In particular I wish to express my gratitude to L. v. Bertalanffy, P. B. Medawar and J. M. Thoday. But they bear no responsibility for my formulations.

L. L. W.

Internal Factors in Evolution

1: Biological Organization

Three basic ignorances mark the scientific scene in our day:

1. The form of a unified theory of physical field-particles.
2. The character of biological organization.
3. The structure of brain-mind processes.

Never before have scientists experienced so vividly the sense of approaching new fundamental unknowns; never have *three* basic themes appeared so ripe with the possibility of major discoveries. For these are each clearly circumscribed problems already subject to experimental and theoretical analysis, and it would be surprising if great advances towards the solution of all of them were not made during this century. But what is perhaps most likely is the discovery of a new mathematical method of analysis and synthesis throwing light simultaneously on all three.

For viewed as theoretical problems the three are closely connected. It is not surprising that in an age óf specialism they should be regarded as independent, but this is far from being the case. These problems all concern complex ordered systems in course of change and any theoretical or mathematical method found appropriate to one of them may aid the solution of the others. It is obvious that a deepened understanding of biological organization should help us to analyze the subtly organized processes of that supremely coordinated organ, the brain-mind. And even if many physical particles appear only at high energies and if organisms are low temperature systems, none the less both require a method for isolating the simple aspects of complex changing systems and there must exist powerful analogies between them.

Of the three the problem of biological organization is in several respects unique and may prove central. It is a problem in three dimensional (3D) space and thus avoids the confusing abstractions of the higher abstract spaces of fundamental physical theory. Next, as the kernel of the urgent "problem of life" it is now of special interest, as is proved by the recent flow of talent towards biophysics. Finally, the present exceptionally rapid accumulation of discoveries in molecular biology and biochemistry has created an unstable situation, for a supersaturation of facts is now awaiting crystallization by the appropriate ideas. Even a hardened skeptic, disillusioned by the relatively slow advance of theoretical biology until recently, must consider that the "problem of life," viewed as the character of biological organization and its relation to physical laws, is near to surrendering part of its secrets. In the past science has made many authentic fundamental discoveries which irreversibly transformed human knowledge. Why not again? And where more likely than here?

The problem of organization may be expressed thus:

How are the differentiated parts and processes of an organism spatially arranged so that the life functions are maintained; how did this arrangement originally come into existence; and how is it re-established in each generation?

When one reflects on the unique importance of a theory of organism for our understanding of the universe and of ourselves, it is puzzling to realize how little is yet known, and how few hypotheses have yet been proposed. For there are only three general attitudes to the science and philosophy of organism.

1. *Aristotle* regarded the organism as a harmonious system in which the form or soul of the organism dominates the parts so that they serve the functions of the whole. After the development of physical science this led to the vitalistic doctrine that some life-principle transcends the quantitative properties of the physical parts. Goethe, keenly interested in morphology, followed Aristotle, stressing the balance and compensation of the parts in the economy of the whole. For him and many after him the *gestalt* of the organic system was more significant than the properties of its separate parts.

2. *Descartes* conceived the organism as a machine

31

composed of separate parts in relative movement. This became the mechanistic doctrine that the global properties of organisms are reducible to the quantitative properties of their smallest material parts. There is no question that this belief has inspired the important recent advances of biophysics and biochemistry.

While the Aristotelian view emphasized the harmonious unity of organisms, it resulted in a relative neglect of *structure* on the part of its adherents. On the other hand the successful Cartesian school, fully occupied with the study of the quantitative properties of small parts, tended to neglect their *ordering* within the living system. As a consequence the central problem of organic morphogenesis has remained very obscure. Indeed the failure to come seriously to grips with plant and animal morphogenesis provides a warning that something may be lacking in the Cartesian way of thinking when it comes to dealing with problems of ordering.

3. The organismic doctrine, developed since around 1920, has attempted to overcome the effete vitalist-mechanist antithesis by emphasizing the organized structural pattern of living systems. An organism is a complex system of relations with a characteristic form of ordering and of change; it is too early for certainty regarding the precise scope of known physical laws; the task for exact biology today is the study of the complex forms of changing order in organisms.

This relatively new school claims no Aristotle or Descartes. It is the collective product of many xxth century minds, who must not all be identified with the doctrine as here expressed. But Whitehead, d'Arcy Thompson, Child, Goldstein, Bertalanffy, Needham, and Woodger are among those who have contributed to it. (All my writings spring from this view of organism.)

One of the most concise formulations of this school comes from Woodger: *"An organism consists of chemical atoms plus ordering relations."* This defines the task of structural biology: to identify and account for the relations which order the chemical atoms. Are these ordering relations covered by already established physical laws, or does their full understanding await a unified theory of physical forces? For it may be that only a unified physical theory can provide a sufficiently reliable basis for a valid theory of so complex a system as an organism, and that both theories will form part of the general theory of complex ordered systems and their changes. If so it is too early for a final estimation of the degree of truth in mechanism and organicism. But organicism appears to be the more fertile approach for the future.

A discrete or unit structure marks all levels of biological organization, and at each level a heterogeneous system of units is subject to a unified ordering. But this ordering is not yet understood.

The structure of organisms has been studied with

great intensity without corresponding advances in the fundamental theory of organization. The discovery of cells, the anatomical and morphogenetic researches of the xixth and xxth centuries, the recent biochemical and cytological work on enzymes and on the ultra structure of cells, the identification of the roles of DNA, RNA, and proteins—all these have greatly advanced our analytical knowledge of stationary structures. But in 1964 we still have no reliable conception of the basic laws of morphogenesis or of the factors which preserve the unity of the organism by determining what processes occur at a definite moment at any given point. The fundamental spatial and temporal laws of the developing organization of local processes are still obscure.

What is missing? Possibly knowledge of certain crucial facts, but certainly a sufficiently clear formulation of the problem. We cannot expect to understand organization until we know what we are looking for *in terms of mathematics*. The use of mathematical standards alone can clarify the aim of a theory of organization.

I have therefore proposed (*Acta Biotheoretica,* 1964) that the purpose of a theory of organization should be made precise by expressing it as a mathematical problem. In order to do this it is of importance that the most appropriate general term should be used to designate the characteristic structural ordering or spatio-temporal *coordination* of parts and processes. *Coordination* is selected as preferable to organization, ordering, correlation, integration, unity, coherence, or balance, which may be better reserved for other purposes.

The mathematical task, as I see it, is to identify the *Coordinative Conditions* (C.C.), i.e. the general algebraic conditions expressing the biological spatio-temporal coordination, the rules of ordering which must be satisfied (to within a threshold) by the internal parts and processes of any cellular organism capable of developing and surviving in some environment. The C.C. are the expression of geometrical, 3D, or perhaps kinematic rules determining the necessary 3D or spatio-temporal network of the relations of the atoms, ions, molecules, organelles, etc. in a viable organism. They are necessary in the individual, not merely statistical over assemblies. The C.C. must cover all the fundamental aspects of the unity of organisms. Thus, to take one example only, the C.C. must define the conditions determining the over-all organization of the chromosomes, of the cell nucleus, indeed of the entire genetic system which determines ontogenetic development.

The C.C. are not merely *morphological,* expressing relative spatial positions and orientations within the ordered system (not in an arbitrary external frame), but *morphogenetic,* representing a one-way tendency towards the development of stationary forms. In other words the C.C. not only define an *invariant characteristic configuration* which tends to persist through all the normal transformations of the system (i.e. through

35

growth, functioning, reproduction, etc.), but also a "dynamic" *tendency towards ordered equilibrium,* i.e. a self-ordering and self-stabilizing process. Thus the C.C. are the mathematical conditions which cover not only the *homeostatic* (feed-back control, etc.) properties of the organism, but also its *one-way* development. Indeed from a logical and mathematical point of view the stationary properties must be secondary consequences, arising under limiting conditions, of a one-way developmental principle. Ontogenetic development is theoretically primary to homeostasis and cyclic function.

The C.C. may be a single set of conditions applying at all levels of structure, from the nucleus of the germ cells to the organism as a whole, or they may be a hierarchy of conditions. They are certainly *strong,* in the sense that they impose a high degree of order, but they are *not maximal,* for they leave many parameters free to vary (fluid regions, flexibility, mobility, learning, successful mutations, etc.).

Though the C.C. may at the present time be easiest to consider in terms of molecular arrangements, they are defined to cover biological coordination at all organizational levels of living material, from the macromolecules such as genes and viruses, through the organelles of cells, tissues, organs, and individuals, and may be extended by analogy to societies of organisms. At each different level the C.C. are paramount in regard to functional activity and survival. But it must be remembered that since we do not yet understand the factors that bridge the levels of organization it is not permissible to assume that parallel or similar conditions hold at all the different levels, though this may be the case.

The C.C. are the general conditions satisfied by all organisms, but they can be met in countless contrasted particular manners—just as an algebraic equation can be satisfied by many particular solutions—corresponding to the hierarchy of phyla, classes, orders, families, genera, species, and variations. The C.C. must reveal why the course of evolution displays a series of structural types following one another in a certain necessary sequence. A mutation that survives development is a change from one particular solution of the C.C. to another, and successful mutations may be graded according to the importance of the change in the mode of coordination. The coding of specificity in DNA, RNA, and protein originates, and is only effective, under the C.C.

The C.C. are to be regarded as the conditions under which the basic laws of complex structured systems permit the emergence and persistence of cellular life, and not as the consequence of past processes of adaptive evolution. For the C.C. are logically prior to the specific adaptive properties of any given type of organism in any particular type of environment; the general conditions which express the possibility of the existence of self-maintaining and reproducing systems had to be met before adaptive evolution could arise; the general fact of organic coordination is, for a theory of organism, a matter prior to changes in the particular mode of co-ordination.

Is this discussion of the C.C. empty conjecture of no present value? Certainly not! Darwin was not prevented from developing a theory of evolution by his ignorance of the mechanism of variation and inheritance; we need not hesitate to outline a general theory of internal selection before the C.C. are known. To any mind fully aware of the importance of ordered structure in the organic realm, the C.C. are less conjectural than the conception of an *ecological niche* in the past history of evolution. Structural biologists may soon know more about the C.C. and their solutions covering particular classes of organisms than evolutionary biologists will ever know about the precise animate and inanimate conditions of the competition for survival, say of one of the earliest birds some 130 million years ago. The conception of an ecological niche is none the less useful; even more so the concept of the C.C., which are already being studied, will soon be subject to exact analysis, and may be given their correct mathematical expression within this century. In no case are all the factors determining reproduction or survival yet fully understood; yet Darwinian theory is undoubtedly useful. The C.C. are unknown, but the conception is already valuable as a guide to clear thinking in an age of structure. We shall see that only with the aid of the conception of the C.C. is it possible to make a fundamental distinction between the two classes of factors, the internal and the external, which together determine the paths of organic evolution.

The discovery of the C.C. may be regarded as the main task of structural biology and of fundamental theoretical biology. For until they have been identified it will be impossible to create a unified developmental and evolutionary biology. In particular the following problems cannot be solved until the C.C. are known:

1. *The mathematical expression of the "unity of organism."* It is the C.C. which make the organism an organic unit, i.e. a system in which local structures and processes are so ordered that a unified result ensues. It is the C.C. which express the unified ordered heterogeneity, the morphological integration of local processes.

2. *The relation of homeostasis and development.* The view that development is theoretically prior to homeostasis can only be fully justified by the discovery of the C.C. It may then become clear why the "regulatory" tendency towards stable end states, inherent in many low temperature systems, becomes fully dominant within organisms.

3. *The relation of local effects to the morphology of the whole.* Established analytical methods tend to neglect morphological relations. The C.C. must reveal the importance of location and orientation within the ordered system, and explain why certain morphological features and metabolic chains have been privileged over others.

Internal Factors in Evolution

Further aspects of the C.C. will be considered in the last chapter. We must now briefly examine the dominant theory of evolution.

2: The Synthetic Theory of Evolution

The theory of evolution by natural selection, which became dominant and reached a certain standard of completion between 1930 and 1950, has been called the "synthetic" theory because it is a synthesis of Darwin, Mendel, mutations, and modern statistical methods. This magnificent fusion of originally separate ideas is a cooperative achievement of many workers, among whom Dobzhansky, Fisher, Ford, Haldane, Huxley, Rensch, Simpson, Stebbins, and Wright contributed much. By 1950 this theory had been presented in a number of comprehensive surveys and it formed a dominant orthodoxy in the sense that there was a high degree of agreement between its leading exponents and that there was no competitive theory. This situation still prevails today. Such skeptics as remain have no alternative to offer. Thus the synthetic theory may be regarded as a splendid culmination of a hundred years of evolutionary and genetic studies, and its character as an integration of separate ideas heightens its prestige. No unprejudiced mind can fail to agree that the theory describes factors which must have played a great part in the past evolution of living forms.

It seemed that biology had at last produced a comprehensive theory as imposing as those of physics. Yet if the history of physics is taken as the norm, evolutionary biologists might have asked themselves what revolution like those of physics might lie ahead of the synthetic theory, perhaps generalizing it to cover an even wider range of phenomena. If they had admitted this possibility it would have been wiser to claim that the theory described *one* major mechanism of evolution, but that there was no evidence that it was the only one. The synthetic theory was a necessary but not a sufficient theory of the directive factors in evolution. However, nearly all the leading exponents of the theory presented it as absolute: adaptive selection, substantially as outlined by Darwin, was *the* mechanism by which adaptively undirected mutations resulted in the progressive evolution of living forms.

A theory of selection must cover two aspects: the production of variation and the choice of variants. In the synthetic theory the ultimate source of phylogenetic change is the occurrence in the germ cells of small discrete mutations subject to recombination, and chromosome aberrations, which are transmitted. These mutations may be changes in genes, in chromosomes, or, as recently discovered, in the cytoplasm or cortex of the zygote, in which case they probably are not Mendelian.

They are conceived as adaptively haphazard, i.e. they are assumed not to possess any *positive* correlation with environmental conditions or with the adaptive properties of the corresponding phenotype, and therefore *cannot themselves constitute a directive factor in phylogeny.*

It is also assumed that the reproductive rate is *always* so high, and the environment so restricted, that "many more are born than can possibly survive" (Darwin), so that there is always competition for an ecological niche, and of any pair of forms it is therefore the one which is better adapted to this niche that reproduces more. Thus allowing for isolation, environmental changes, etc., a slow cumulative evolutionary advance in fitness for life and reproduction can be produced. The sole directive factor in phylogeny is the selection of the better adapted forms.

That is the logical skeleton of the argument. The successful evolution of one type as against another is due to a difference in reproductive efficiency which is taken as evidence of a selective process acting on mutants in terms of their adaptive capacities in relation to particular environments. This process both maintains successful types (stasigenesis) and in suitable circumstances results in the emergence of better organized, more efficient types (anagenesis).

Many secondary considerations are also involved and the application of the theory to particular problems has resulted in elaborate discussions. But the central argument rests on these conceptions: haphazard mutations and recombinations which are preserved; high reproduction rate, inadequate living space, high death rate, and differential reproduction; isolation of populations, evolutionary advance, splitting, and stabilization.

It has been suggested that the reproductive differential on which the mathematical theory rests should be divided into two parts: a true reproductive (sexual) differential, and a survival differential, but this division is not generally accepted. The crucial characteristic of the synthetic theory is *a net reproductive differential of genetic variants, regarded as due to differences of adaptive fitness which are thus the causes of phylogenetic change.* This may be taken as a definition of Darwinian, adaptive, or *external selection.*

Some exponents of the theory have sought to include within its scope "anything which produces a genetic differential in reproduction" (Simpson), but this, as we shall see, begs an important question and conceals a crucial distinction. For modes of selection which are *not* determined by adaptive efficiency may none the less produce a survival differential of hypothetically possible genotypes, effectively equivalent to directed mutations. It will here be argued that such modes of selection should not be regarded as being of Darwinian type, since they involve basically distinct criteria. To this we shall return in the next chapter.

The Synthetic Theory of Evolution

The synthetic theory was not so complete as to stifle all advance. The reverse has been the case; the theory has stimulated and absorbed many new ideas during the period of its dominance. For example, it has become clear that it is not the single mutated gene which is selected, but the phenotypic consequences of the entire harmonious genotype in which must be included any transmitted features of the cytoplasm and cortex of the female gamete. Again, it has been suggested that in small populations a "drift" may occur producing characters which do not possess adaptive value. These are two examples of many recent ramifications of the theory.

The theory has been subjected to various criticisms of which the following are relevant:

1. Owing partly to the absence of any direct test for adaptive fitness, the theory appeared to some to be capable of being adjusted to account for every conceivable kind of evolutionary change, not merely those which have actually occurred. This criticism was partially met by the mathematical developments of the theory, some of which have been experimentally tested.

2. There is no adequate theory of variation. With the growth of molecular biology it became evident that the meaning of "random" mutations required further analysis, particularly as most mutations of an ordered system must be injurious. The crucial issue now is: Are the mutated genotypes *whose consequences are submitted to external selection* random or not in relation to the sub-

sequent directions of evolutionary change? It is here asserted that they are almost certainly *not,* having been already subjected to internal selection.

3. No major favorable mutations, i.e. other than minor modifications of established types, have yet been identified or studied. This remains a serious weakness of the observational basis of the theory. Can the synthetic theory be held to account for one of the most important features of the evolutionary past: the emergence of entirely new more highly differentiated types, until the corresponding favorable mutations have been identified? In spite of considerable discussion there is as yet no general agreement on this issue, and it seems that there cannot be until such macromutations or their equivalent are better understood.

4. Some embryologists have held that the synthetic theory cannot be regarded as definitive until it has been combined with a theory of ontogeny. Not only are ontogeny and phylogeny in many respects related but ontogeny is theoretically primary. This is embarrassing for evolutionary theorists, as no adequate theory of ontogeny is yet in sight. Moreover, those who are concerned with these or similar problems (e.g. Waddington, Lindegren) are being led to quasi-Lamarckian ideas. A survey of the recent literature on these themes suggests that in the absence of new clarifying principles verbal discussions are in danger of becoming overelaborate and inconclusive. The needed principles may perhaps arise from the side of molecular biology.

The Synthetic Theory of Evolution

5. Apart from these difficulties an independent mind may question whether too grand and comprehensive a theory claiming to cover distant past history has not been built on relatively thin evidence, however attractive and powerful its ideas. For example, something is surely wrong when population statisticians discard the biologically significant concept of *fitness* and treat as an adequate substitute a formal statistical parameter: differential reproduction; or when the applications of a theory to largely hypothetical historical situations require lengthy verbal analyses in terms of high conceptual abstractions beyond observational control. There is little doubt that mathematics or mathematical logic must be brought in to help and that it should be a form of mathematics directly related to particular biological situations, not merely to the statistics of populations.

Yet these issues do not prejudice the achievement, provided that the theory is correctly interpreted and its conclusions are stated with scientific caution. The synthetic theory is an advance of the first magnitude and its marginal difficulties are a sign of its fertility in opening up new vistas of thought and observation.

This judgement is not affected even if, as is often the case, it turns out that hidden redundant assumptions have been made, that formulations have often been imperfect, and that some claims were expressed too generally.

The exponents of so great a theory should be willing to discover that it only holds one aspect of the truth. In

Internal Factors in Evolution

any case the valid parts of the theory remain as an indispensable preliminary to the next step: the discovery of a second class of factors that have played an important role in directing the evolution of living forms. For any structural thinker must conclude that Darwinian adaptive selection has not been the only directive agency guiding the evolving forms of life along the paths which they actually followed. The nature of life, the structural character of organisms, has itself imposed basic restrictions on the changes that are permissible. It has only been possible to neglect this natural conclusion because so little is yet known about this internal coordination which is "life." But once explicitly considered, this internal selection is seen to be just as natural and inevitable as Darwin's "natural selection."

3: Internal Factors

Anyone aware of the pervasive importance of fine structure, having in mind, for example, the role that snow crystals have played in the evolution of mountains and rivers, would tend to assume that the fine structure of organisms must have played a similar part in the evolution of the forms of life. He would expect that the synthetic theory only covered the macro aspects of organic evolution and that a future micro theory would be necessary to complete it.

Viewed at any level, from organ and tissue down to cell and component molecules, the organism is a highly ordered system. This is true both of structures and processes. The macroscopic organs and their functions fit together. Similarly the basic structures, such as the chromosomes, form a coherent pattern and undergo collective motions and transformations displaying a high degree of spatial and temporal coordination. This coordination is pervasive at all levels. The ultra structure of the living cell is an intricate differentiated network undergoing global pulsations and transformations under some law of ordering which preserves the unity of the

49

system. These are clumsy words beside the elegant harmony which they seek to describe.

Our knowledge of certain stationary aspects of this ordering has recently become more precise at the molecular level. The chromosomes contain helices of DNA embedded in a protein matrix, and mutations consist either of alterations in the sequence of chemical units making up the DNA, or of transmitted changes in the cytoplasm or the cell cortex of the female gamete.

It is obvious that entirely arbitrary changes in this genetic structure will not be physically, chemically, or functionally stable. Only those changes which result in a mutated system that satisfies certain stringent physical, chemical, and functional conditions will be able to survive the complex chromosomal, nuclear, and cellular activities involved in the processes of cell division, growth, and function.

"The struggle for survival of mutations begins at the moment mutation occurs" (Whyte, 1960), but this is a "struggle" to conform, not to compete.

The greater we assume the degree of order in the cell, the more stringent these conditions will be, and they certainly must be severe, that is highly restrictive of the permissible changes. For example, at the chemical level only a limited number of stable macromolecular arrangements are possible—say in the pattern of atoms constituting a gene—and these in turn must conform to unknown principles of over-all ordering which must be satisfied if the various chromosomal activities are to be carried through without confusion. Only certain changes in the composition of the DNA and the arrangement of the chromosomes will be possible without loss of the faculty of replication or of other chromosomal functions. All the parts of the genetic system must be adapted to one another so that they work together. This sets limits to the potential of variation. Moreover the conditions of ontogenetic development set further restrictions, as Waddington has emphasized. All these restrictions fall under the C.C.

Thus the number of permissible changes at each gene locus is limited, and the conditions which determine the unity of the entire genotype must place further restrictions on the potential for variation. If fully arbitrary mutations were permitted there would be a continuum of possibilities. As it is, only a discrete spectrum of variations is permitted by the C.C. Organic "quantum" conditions restrict the potential variations.

The mutated genotype must *first* satisfy the C.C., and *second* not be so different from the unmutated genotype as to prejudice a successful transition. Since the previous coordinated system has been discarded, a new and sufficiently accessible and stable coordination must replace it, if the mutated system is to develop successfully. Only certain classes of mutations will permit the mutated system to satisfy the C.C. in such a manner as to permit a successful transition.

Internal Factors in Evolution

To employ a useful analogy: Not only the general differential equations of life must still be satisfied, but the new particular solutions of the equations which represent the mutated system must not involve too difficult a transition from the previous solution. Mutational "selection rules" must determine the permissible transitions and so restrict the avenues of evolutionary change.

This argument applies at every level and stage of development, from fertilization through the first cell divisions until the developing organism meets the challenges of survival and reproduction.

The structural approach thus leads to the following conclusion: *The conditions of biological organization restrict to a finite discrete spectrum the possible avenues of evolutionary change from a given starting point. The nature of life limits its variation and is one factor directing phylogeny.* Moreover, it is possible that in certain rare but important circumstances, which will be considered later, these internal conditions by themselves fully determine the evolutionary changes that occur.

This argument implies that even if the premutational factors (radiative or thermal fluctuations, or mechanical accidents) are haphazard, i.e. possess no correlation with the directions of phylogeny, those mutated genotypes which survive development must have satisfied criteria which impose restrictions on the permissible paths of evolutionary change. The mutations whose consequences reach the Darwinian test are not necessarily random in relation to phylogeny for they have already been sifted by an internal selection process.

Thus the position has drastically altered since, say, 1910. Then it was appropriate to assume that mutations were adaptively random. Now the term "random" is seen to be not merely vague, but positively misleading. It neglects the high degree of order in organisms and should only be employed in special senses where statistical tests establish the absence of some particular correlation. Further, the long association of the term with adaptive randomness conceals the crucial fact that *mutated genotypes which have passed internal selection will in certain circumstances have a positive correlation with the directions of evolution.*

It must be stressed that no Lamarckian features enter the present argument, nor is any kind of somatic selection involved since we are solely concerned with mutations occurring in the germ cells.

Whereas in the synthetic theory the *sole directive factor* in phylogeny is asserted to be Darwinian external selection acting on mature forms in terms of their reproductive efficiency in particular environments, now a *second directive factor* has emerged: the prior requirement of high internal order expressed in the C.C. Life can only evolve within broad avenues determined by its own structural nature; the actual paths followed within these avenues are then determined by Darwinian external selection. It is the prior internal selective process,

53

Internal Factors in Evolution

rather than Darwinian selection, which ensures that only well integrated genotypes survive. Internal selection has hitherto been neglected for two main reasons: little has been known about it, and it is probably mainly associated with those changes in general organization about which we are also ignorant.

These are the provisional conclusions to which one is inescapably led by taking seriously the high internal order of organisms and in particular of genotypes.

The mutated genotype which satisfies the C.C. develops successfully and proceeds to the second test of adaptive selection. But how are those genotypes which do not satisfy the C.C. eliminated? This may, it seems, occur in at least three ways:

1. An inappropriate molecular configuration may be forced back to its original state by a *return* mutation.

2. The mutated or disturbed configuration may, in the course of the earliest chromosomal activities, be molded into a *new* configuration which does satisfy the C.C. This hypothetical prefunctional adjustment of a non-conforming disturbed genotype will be called its *reformation*. Though there is as yet no direct evidence for this effect, it is a type of assimilation and is similar to the "adjustment" of mutations which has been discussed in the literature, and it appears to be a probable consequence of the operation of powerful C.C. A complex ordered system in the course of its functional activities is likely under known physical principles (e.g.

the tendency towards configurations of minimal potential energy) to possess the power to adjust parts that do not properly conform. This is the self-repairing property of organisms viewed at the basic structural level, but with this important difference: a *new* functional configuration results. As a possible source of favorable mutations this mechanism is of great importance and requires more detailed consideration.

Medawar (1960) has suggested that there is in systems of polymers a "repetitiousness" or tendency towards elaboration, so that evermore complex sets of genetical instructions are offered for trial, and that this may provide a basis for advance in complexity. This idea receives support from physical theory for, as Bachelard (1953) has stressed, there is implicit in quantum mechanics when applied to complex systems a *structuring tendency,* for example a tendency for macromolecules to combine or arrange themselves to form more complex forms of ordering. Quantum chemistry reveals this structuring tendency as an activity which tends to fill space in accordance with continually more complex symmetry conditions.

In organic systems undergoing collective functional cycles these new forms of ordering will necessarily be *unified* in the sense that the system as a whole must be capable of such global cycles. This may imply that mutations or disturbances which are initially arbitrary, i.e. bear no relation to the functional system and are therefore likely to be harmful, are adaptively modified or assimilated to the system of which they are part, as has been suggested by Vandel (1963). For example, the functional processes undergone by the nucleus may tend to mold arbitrary and unstable molecular arrangements into more complex but also more unified and stable patterns.

This tendency towards the formation of more complex unified patterns does not imply any obscure vitalistic factor, since in appropriate circumstances it can be the direct result of the tendency towards arrangements of minimal potential energy. Thus the potential energy principle can, in complex low temperature quantum mechanical systems, produce a structuring or formative tendency which, under certain conditions, will shape the genetic system towards novel, stable, unified arrangements. Arbitrary changes in the genetic system may thus be reformed into favorable mutations satisfying the C.C.

3. Such return mutations and reformation will only be possible when certain critical thresholds have not been passed. In other cases the defective genotype will result in the death of the resulting cells, tissues, or organisms.

These lethal mutations have been known for several decades. But it has only recently become evident that the existence of such lethal mutations, if they cannot be compensated by a change in the environment, implies the operation of a second radically distinct kind of selective process, one in which the criterion is not a matter of competition but of coordination. Deleterious or lethal mutations, expressing a failure of coordination and not merely the failure to produce some metabolite which the environment might supply, involve a new type of selection.

It has been widely observed, by Waddington and others, that developmental homeostasis, or a well-knit system of the canalization of development, severely narrows down the evolutionary potential. But it is important to go further and to recognize that this is the expression of a new kind of selective and directive factor leading phylogeny along certain avenues and prohibiting certain other hypothetically conceivable ones.

An attempt has been made by some of those who have considered the direct action of internal factors to distinguish between (i) alterations in the genetic material which render it incapable of taking part in the earliest cytological processes such as chromosome duplication and separation in mitosis; and (ii) mutations which result in developmental failure at some later stage. When more knowledge is available this subordinate division may be useful for some purposes. But at present it is appropriate to treat together all selective processes which depend on internal factors and involve the C.C. Thus we define *"internal selection" as the restriction of the directions of evolutionary change by internal organiza-*

tional factors, i.e. selective processes acting directly on the early consequences of the genotype which ensure that the C.C. are satisfied by all mutated types that survive up to the point at which Darwinian external selection enters. (This internal selection in cellular organisms is to be distinguished from Oparin's (1961) preorganic "chemical selection," e.g. of stabilizing factors in a coacervate droplet.)

A naturalist, or anyone else unaccustomed to structural thinking, may regard this definition as obscure. But it is no more so, and in some respects much more reliable, than the *ecological niche* (or adaptive zone) for, as we have seen, this has never been given more than a very general and vague specification. Both terms, the niche and the C.C., are useful in framing hypotheses without which a potentially comprehensive theory of evolution cannot be formulated.

Some who have considered the operation of internal factors have argued that what is here called internal selection is merely Darwinian selection extended to cover the internal environment. It is suggested that any attempt to distinguish the two fails because on a truly relational view in which interactions are recognized as pervasive there is no distinction between the internal and external environments. "Natural selection" occurs at all levels, everywhere.

Nothing could better display the harm of intellectual fashion. It is good style today to equate the internal and external environments; therefore there can be only one universal type of natural selection. Thus contemporary habit tends to inhibit the recognition of a distinction which is indispensable to clear thinking. For the internal and external environments are not alike; life is not a vague property diffused throughout the universe; the nature of life lies partly in the fact that it is concentrated in units whose interior is subject to an ordering principle which does not apply to the inanimate spaces between organisms. These units of organization do not lose their distinguishing form of order because they are in perpetual give and take with the environment; the C.C. do not operate through open spaces, they are only effective inside boundaries. Thus the two kinds of selection can and should be distinguished.

Darwinian adaptive selection is a relative, statistical, external process effective between pairs of individuals or populations, a matter of degree determined only by comparative fitness and frequencies in some particular environment. Internal selection is an intrinsic, usually all-or-none process operating within single individuals, determined by observing the history of one organism in the most favorable environment. Adaptive selection depends on a high death rate and expresses competition; internal selection rests on coordination within individuals.

The evidence for internal selection is not a relative rate of increase in a given environment, but the direct structural observation of the successful (or unsuccessful) internal coordination of a single mutated organism.

The distinction could not be logically more definite or biologically more significant. Adaptive selection requires one kind of reasoning and mathematics, internal selection another.

The use of the C.C. to discriminate internal from external selection may enable the two to be distinguished even for unicellular organisms in intimate relation to their chemical environment. For structural observations can in principle discriminate between the failure of a mutation to conform to the C.C. (even in the most favorable chemical environment) and the adaptive failure of an organism to reproduce, in some particular environment.

When a new idea, far-reaching in its implications, is first considered, the mind oscillates in discomfort between novel vistas and long-established tracks of thought. Stable clarity regarding a new fundamental idea is not possible until many bordering problems have been reconsidered from the fresh point of view. The idea has to be applied in many directions, its distinction from earlier ideas examined in its many aspects, and misunderstandings removed by developing improved formulations. This is a slow and troublesome process even for those whose energies and time are free, but it can be assisted by two techniques: discussions with colleagues and systematic comparison with earlier ideas.

One purpose of this exposition is to try to bring to the busy reader the benefit of these two procedures, so that if his mind is open and this essay adequate to its theme, he may see the idea clearly and form his own judgment as to whether it is timely and fertile, particularly in relation to his own problems if he is a biologist.

But while this exposition is personal, the idea itself has a collective origin. I would not have spent time on it if the fact that others had come to it simultaneously and from contrasted approaches had not convinced me that the time has arrived when it can be made fertile, and that it is therefore worth while trying to get it clear, or as clear as is possible before going into its biochemical, genetic, and other special applications.

If this new school of structural evolutionary thought is not entirely mistaken, there is no doubt of the need for an exposition such as this. Take one example from many, chosen to establish this point and not implying any lack of respect for a thinker who is a master in his field. The latest, most authoritative, and detailed discussion of the synthetic theory is E. Mayr's *Animal Species and Evolution* (1963), reviewed by J. S. Huxley as a "magistral work." Here it is repeatedly stated that "selection acts on phenotypes" (the context showing that developed organisms are meant) and that Darwinian selection is *the* directive factor in phylogeny. This is incautious and unscientific. For it certainly has not

Internal Factors in Evolution

yet been established that there is no selective process acting in terms of internal coordination. An assumption has unconsciously been made which in the present argument is certainly not established and is very probably incorrect. Even if the role of internal factors is disputed, they should not be left undiscussed in a scientific analysis ten years after four thinkers have declared their conviction that internal factors may sometimes be important. No future survey of evolutionary theory can be regarded as competent which fails to discuss internal selection.

4: Internal Factors (Cont.)

In the long perspective of the history of ideas the synthetic theory is very recent, too new for a school of younger minds to have matured capable of identifying all its unconscious hidden assumptions. One such assumption has recently become evident and is now seen to be questionable: that a fully reliable or definitive theory of organic evolution is possible *before* ontogenetic development is understood, and this means structurally. This is from one point of view a strange assumption, for logically and theoretically ontogenesis is prior to phylogenesis, which is a sequence of ontogenetic stories. The growth of an individual can in principle be described without considering the evolution of the forms of life, but evolution is meaningless unless there are individuals which have grown. Ontogenetic development should be understood before historical changes in that development; the structural story of growth underlies the natural history of changes in growth; a life cycle must be present before it can undergo changes, a genetic system must exist before it can mutate. From one point of view, this amounts to no more than saying that the more exact science seeks to become, the more firmly must its principles be grounded in structure; a molecular biology, adjusted to allow for the unity of the organism, must ultimately provide the basis for a universal biology including a unified theory of ontogeny and phylogeny.

The acceptance of a structural approach as an indispensable element in a complete description involves more than a mere change of emphasis, for it corrects definite mistakes. The following statements, found even in some of the best recent texts, are from the present point of view either clearly wrong or likely to be proved so in the near future:

All selection is expressed in population statistics. No! Internal selection is evidenced in the structural history of individuals, and some excluded types never enter population statistics.

All properties or organisms are the consequence of past adaptive selection. No! Many properties, especially those of coordination, may be the result of internal selection.

Any selective process that may occur within an organism is of Darwinian type. No! The reasons have already been stated.

These challenges to orthodoxy are not made for the sake of polemics, but to clarify thinking.

Consider, for instance, the school of population statistics, which starts with the excellent aim of fully objective numerical observations, but sometimes slips into asserting that population figures provide the *only* basis for a theory of evolution. This is wrong. The statistical theory of populations must be complemented by a structural theory of individual ontogenesis and of its influence on phylogeny. This may reinstate a theory of the typology of phenotypes based on a typology of genetic systems and of proteins, etc.

That the conditions of biological organization and development necessarily impose restrictions on the possible lines of evolutionary change is easy to recognize today, at least for those accustomed to structural thinking; but it is the kind of reasoning which is far from reliable until the time has arrived when its implications begin to be understood. Countless naturalists before Darwin must have observed that the better adapted types tend to survive without generalizing this observation into a universal principle that because more are always born than survive there is a *general* tendency towards greater fitness as regards hereditary capacities. In the same way few have as yet inferred from the occasional presence of lethal or deleterious mutations which cannot be compensated by changes in the environment the operation of a new and potentially general mechanism of selection. Indeed at first sight it is far from obvious what the combined operation of two distinct methods of selection implies.

For example, the position may be—and certain arguments support this view—that internal selection *always* operates, i.e. sifts the mutated genotypes, allowing only some to survive development, but that external selection, though normally operating, fails to do so in certain circumstances, for example when it is *not* the case that "many more individuals are born than can possibly survive" or where the genotype changes without change of phenotype. If we grant, for the moment, that these conditions may sometimes occur in the wild, then it follows that during such periods internal selection alone may have been responsible for determining the avenues followed in the past history of evolution. Groups of organisms may have emerged and in special situations or during restricted periods evolved along certain avenues as a result of internal selection alone.

We may here neglect the (here) relatively unimportant case of genetic evolution without phenotypic change, and concentrate on the question: When would a population be least subject to restrictions on its multiplication? When might most of those born survive to reproduce? The answer can only be: when the population is relatively small and occupies a new ecological niche, previously empty; that is, when a small number of individuals of a basically new type are present. These conditions are best met when a major genotypic change is occurring which leads, say, to a new phylum, order, or class, or when living cells are first emerging. For until the numbers of this new type are sufficient to fill the corresponding niche there will be no restriction on its multiplication and no Darwinian type selection will occur. Darwin pointed out that "the greatest amount of life can be supported by great diversification of structure." This implies that forms with an entirely new type of structure may find awaiting them ample space, diet, and security—until in turn new families and species have filled up the free ecological spaces and begun to face a new competitive struggle. Thus Darwin's major assumption of a *universal* restriction on free multiplication is least valid, and may perhaps entirely fail, in special periods of great evolutionary significance.

"Adaptive superiority" is empty of all meaning when an entirely new form emerges and occupies a new adaptive zone, for there is then no competition and no basis for a comparison. It is difficult to avoid the conclusion that the nearer the origin of an entirely new type and the smaller the numbers present, the less important is external selection.

The emphasis placed here on the unified internal ordering of the organism suggests that in certain circumstances a global factor may be present tending to produce many simultaneous minor quasi-mutational changes, forming components of a single major change in organization. As we have seen, non-conforming small mutations may be reformed to conform to the over-all transformation. This would produce quantum evolution, its discrete or stepped character arising, not from a change of habitat, but from a discontinuity in the range of internally permissible structural arrangements. The discrete spectrum of permissible genetic variation produces a discrete manifold of phenotypic forms.

Thus we are led to consider the hypothesis that internal selection may be of decisive importance at periods of macrogenesis when the more fundamental changes in the evolution of organization occur. At such times of genetic revolution, when relatively few founders are able to originate a wholly new type which selects its own environment, the normal rules of population statistics fail because it is no longer true that many more are born than survive and because the numbers are small. If myriads of species and a stupendous number of individual organisms have emerged from relatively few origins, then it is necessary that at some periods some species must have displayed a high rate of increase. During those periods internal selection is likely to have played an important role in determining the actual direction of advance out of all the hypothetically possible lines.

This suggests that the major features of the evolution of genetic systems (extensively discussed by Darlington) may have been partly guided by internal selection. Moreover, this macrogenesis, being a matter of the general mode of organization, rather than of particular catalysts, may well have been determined by the relation of the over-all arrangement of the chromosomes to the cytological or cortical hereditary factors, not by particular genes. In turning our attention to coordinational properties we are led to pay more regard to global hereditary factors and less to point or gene mutations.

With this emphasis in mind let us consider the possibility that all genotypes face internal selection, but only some external selection. Those mutated genotypes which do *not* pass internal selection fall into at least three classes:

1. Those which kill the individual before maturity or render it infertile.
2. Those which are eliminated by a return mutation at an early stage.
3. Those which are changed by reformation into a different genotype which does pass internal selection.

Of those mutated genotypes which pass internal selection a special class (just discussed) escape external selection, while the remainder undergo external selection, and are statistically sifted by it. These choices are set out in the hypothetical scheme on p. 70, the purpose of which is to facilitate thinking. As shown in this diagram, the direction of evolution may be (i) coordinative alone, or (ii) coordinative *and* adaptive. It is never adaptive alone.

This is a conjectural scheme, but no further from observation than many of the abstract and elaborate arguments regarding drift, mega-evolution, etc., which

Internal Factors in Evolution

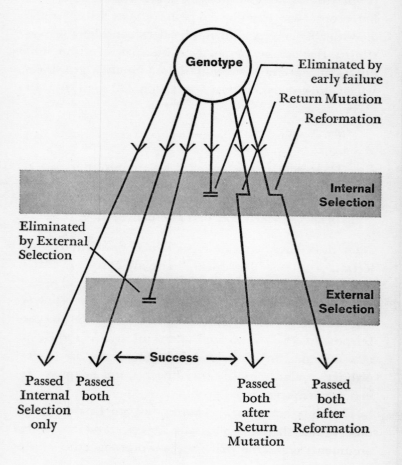

Combination of Internal and External Selection

Genotype

Eliminated by
early failure

Return Mutation

Reformation

Internal
Selection

Eliminated
by External
Selection

External
Selection

← **Success** →

Passed
Internal
Selection
only

Passed
both

Passed
both
after
Return
Mutation

Passed
both
after
Reformation

fill pages of the current surveys of evolutionary theory.

The present generalized argument has great potential power. Einstein said that "once one has sufficiently strong formal hypotheses, one requires little knowledge of the facts to set up a theory." If the formal assumption is happily chosen, valid specific consequences may follow. The same is true in biology. From Darwin's daring assumption of universal restrictions on the rate of multiplication (one which we now recognize may not always be valid) the results of his theory followed, when certain supplementary principles were combined with it. *Darwin's ignorance regarding variation did not matter.*

The corresponding formal assumption adopted here is that *organisms are highly ordered structured systems with which only certain classes of genotypes are compatible.* Though we are ignorant of much detail there follows from this assumption the operation of a new mechanism of phylogenetic change which, combined with that of the synthetic theory, may lead towards a comprehensive and unified theory of evolution.

It is instructive to examine afresh the arguments in the opening chapters of R. A. Fisher's *Genetic Theory of Natural Selection* (1930). Reasoning which appeared to him comprehensive and compelling is that no longer, for he considered only mutations of the kind already observed: those producing quantitative variations. His argument that no directive factors other than the natural selection of undirected mutations can have been at work fails, because he did not consider among the hypothetical directive agencies the possible influence of structural organization. At no point does he even mention internal structural factors. This may not be surprising in a work published in 1930, but nothing proves more clearly the need for a reconsideration of the fundamentals of evolutionary theory than an examination of Fisher's basic assumptions.

There is a final point of some importance. If the process of reformation of disturbed genotypes (or some equivalent) ever occurs, then at such moments internal selection plays a direct and positive role in shaping the avenues of evolution. Internal selection would here, in a certain degree, directly *determine* the variations resulting in phylogeny (provided they pass the Darwinian test), rather than merely *select* from given biologically arbitrary variations. Here internal selection becomes internal determination, and phylogeny the direct result of internal factors. This would scarcely surprise the molecular biologist, for, as we have seen, in certain circumstances there may be powerful physical and chemical forces tending to shape a disturbed genotype into a new functionally stable form. Indeed it may be that once a disturbance to the organization of the genetic system is stable enough to be regarded as a hereditable mutation it has already been molded to conform to the C.C.

The main steps of my argument may be summarized thus:

1. Organisms are highly coordinated structures.

2. Only certain avenues of change are compatible with their conditions of coordination.

3. The formative and selective action of these internal conditions is theoretically and empirically different from that of Darwinian selection.

4. Mutations in the mode of coordination of the genetic system lie outside the scope of the classical arguments purporting to show that natural selection is the only directive agency.

5. The coordinative conditions constitute a second directive agency.

Our first study of the idea is now complete. But an historical survey will help further to clarify the idea, after which some difficulties will be examined.

5: History of the Idea

There are several advantages in considering the historical background of a new idea. An historical survey can throw light on the precise meaning of the idea by showing what it has in common with, and where it differs from, earlier similar ideas. It may lessen the influence of fashion by revealing a long heritage, and reduce any merely personal features and heighten objectivity by emphasizing the collective and convergent features in the development of a major idea.

In the history of the conception of internal selection there have been two distinct periods: 1880–1910, when Roux, Weismann, and Wallace considered and accepted a similar but different idea; and 1940 to date, when it has appeared in a new form and been given increasingly precise formulation in a changed context by several independent thinkers.

1880–1910. In 1881 Wilhelm Roux extended the Darwinian conception of the struggle of individual organisms to a struggle of parts within the organism, using as the title of an essay: *Der Kampf der Teile im Organismus.* Roux conceived this struggle as proceeding at

75

Internal Factors in Evolution

every level from molecules and cells to tissues and organs. The competition of the parts for nutrition to permit them to grow determines the process of differentiation. This was a truly Darwinian conception, Darwin's mode of thought being here extended without modification to the internal environment.

In 1896 August Weismann, in his essay *On Germinal Selection,* adopted Roux's idea, called it Intra-selection, and gave it a new emphasis by applying it to the germ cells and the "determinants" which stabilize hereditary characters. Weismann assumed that the primary variations were accidental, and claimed that his germinal selection provided an internal mechanism which compels the variations to increase in a definite direction, new characters thus arising inevitably from internal causes. He saw evolution proceeding at three levels: by the individual selection of Darwin and Wallace, the cellular and tissue selection of Roux, and his own germinal selection, which he regarded as a source of directed variations.

If one remembers that during this period there was no accepted theory of variations (their character, source, structural basis, stability, etc.), Weismann's analysis is remarkably clear, and his ideas were influential. A. R. Wallace, who survived Darwin by thirty-one years, broadly accepted Weismann's doctrine of germinal selection in his *World of Life* (1911): "There are therefore both an internal and an external struggle for existence affecting all the special parts—organs, ornaments, etc.—of every living thing."

While it is of interest to look back and to observe the speculative energy with which Roux and Weismann sought to extend the Darwinian conception to a new field of which little was known at the time, such intraselection must not be considered as an anticipation of the internal selection with which we are here concerned. For Roux and Weismann were both thinking of a *struggle* between parts, i.e. a competition for space and nutrition in which the relative success of one part implied the relative failure of others. Nothing could be more different from an internal selection in which the criterion is the capacity of all the parts to fit together to form a single *coordinated functioning unit*. It is only when one has meditated on Weismann's idea and its implications, and contrasted it with internal selection, that one can fully appreciate how far-reaching is the novelty of the latter, essentially xxth century idea. For the competitive and random emphasis of the xixth century in certain fields is only now, as our century matures, being balanced by an adequate complementary attention to coordination. Intrinsic fitness for coordination is now supplementing relative fitness for competition. Only when the internal functional coordination for some reason fails does the Roux-Weismann competitive struggle of the parts become a reality, and then we have cancer, not ordered differentiative growth or evolution.

Internal Factors in Evolution

During the XIXth century there was insufficient knowledge to guide speculation about the relative importance of external and internal factors, and it is not surprising that in 1888 T. H. Huxley considered it "quite conceivable that every species tends to produce varieties of a limited number and kind and that the effect of natural selection is to favour the development of some of these, while it opposes the development of others, along their predetermined line of modification." This was, in effect, a partial anticipation of internal selection.

But at that time there was no evidence for such internal determination and the dominant view since 1910 was expressed by T. H. Morgan (1919) who argued that it was "thinkable" but "not probable" that "the direction of mutation was given in the constitution of the genes." It was natural that after Darwin, Mendel, and the discovery of mutations, the emphasis was first placed on the external selection of adaptively undirected mutations; it was Darwinian selection acting over long periods that imposed order and directed sequence on what was otherwise haphazard and accidental. External selection alone was responsible for the directive aspects of the evolution of the forms of life.

None the less a modified form of the principle rejected by Morgan was meantime being silently introduced into evolutionary theory by the students of lethal and sublethal mutations, who took for granted that deleterious mutations and the resulting mutated genotypes might be eliminated during ontogeny, with the implicit corollary that developmentally viable mutated genotypes are restricted to classes determined by the genetic system itself and by the developmental processes which it induces. Even if mutations are originally un-directed, the concept of developmentally deleterious mutations implies that some mutations are not deleterious, i.e. the operation of an endogenous sifting process with its own criterion at work, as well as, and usually prior to, external selection. This inference is justified when the deleterious mutations are such that they would remain deleterious in any conceivable environment.

1940 onwards. Around 1930–40 three influences were preparing the way for reconsideration of the role of internal factors: (i) It was recognized that external selection operated not on single genes, but on the entire genetic system as a working unit, and system effects began to be investigated. (ii) It became increasingly obvious, with the development of structural biochemistry, that the organism constitutes a highly organized "environment" for genes and their mutations. (iii) Closer attention began to be paid to the relations of ontogenetic development to phylogeny, by Waddington and others. Thus by 1940 the background was favorable to the examination of internal factors.

None the less between 1920 and 1940 relatively few biologists explicitly considered in a general context the possibility that the constitution of an organism may not only result in the elimination of deleterious mutations but also set restrictions to the directions of evolutionary change open to its descendants, in contradiction to the assumption of the synthetic theory that external selection is the only directive agency in evolution. J. C. Smuts (1926) asked, "Are [variations] all individually selected before they have any survival value or strength?" "It is thus the organism, as a whole which in the first instance 'selects' the variation." But this was a mere philosophical conjecture, and the attention of most working biologists concerned with the causes of evolution was concentrated on the external adaptive factors, and members of the dominant school only began to re-examine the status of internal factors after 1940.

It is not easy to be just in selecting examples and to avoid reading present conceptions into earlier documents, but the contexts from which the following quotations are taken support the view that their authors were moving in the direction of internal selection. I have excluded references where the same author has subsequently shown that he had no sense of the possible importance of internal selection as a mechanism distinct from the Darwinian.

H. T. Pledge (1939). ". . . the gene-complex is an environment for mutations in the organism."

T. Dobzhansky (1941). ". . . it would follow that the evolutionary courses of races and species differing in gene arrangement are likely to diverge owing to modification of the rates and possibly of the directions of mutations in certain genes."

C. Stern (1943). (Observations on mutations suggested that) "new points of attack for selective forces would originate. . . . It can hardly be estimated how much of such concealed evolution (i.e. internal selection leading to different genotypes, phenotypically alike) due to the kind of 'germinal selection' described, is taking place at any time."

The above must be regarded as tentative suggestions, lacking any clear indication of the operation of an entirely new mechanism of organic evolution. But between 1949 and 1952 *four* writers expressed this conception with greater or lesser clarity. This is an interesting example, not merely of simultaneity showing that the time for the idea had arrived, but also of Nietzsche's observation that the greatest ideas often come into the world "with the feet of doves." For apparently no one noticed until some ten years later that around 1950 a major turning point in biological thought had been reached.

Here are four unmistakable statements of the new principle, two from working geneticists and two from thinkers concerned with the philosophy of biology:

H. Spurway (1949). "A group of related organisms is characterized by similar possibilities of mutation—these possibilities of mutation determine the evolutionary possibilities of the group . . . it suggests a specific control of mutability in excess of anything we know so far. . . . A given species, family, or class mutate more readily towards certain phenotypes than others . . . the mutation spectrum of a group may . . . determine its possibilities of evolution." *L. L. Whyte* (1949). (See passage quoted in Preface.)

L. v. Bertalanffy (1952). "Thus the changes undergone by organisms in the course of evolution do not appear to be completely fortuitous and accidental; rather they are restricted, first by the variations possible in the genes, secondly, by those possible in development, that is, in the action of the genic system, thirdly, by general law of organization."

A. Lima-de-Faria (1952, 1954, 1956, 1962). Certain evidence favors the view that "selection should take place not only at the organism level but at the chromosome level. . . . The genotype of an organism evolves under defined conditions" (1954). He suggested (1956) that "the constitution and organized pattern of a chromosome are the prime determinants of its evolutionary trend and that *the genotype of an organism evolves chiefly under conditions defined by the constitution and organization of its chromosomal components.*"

These passages imply that the genetic system of a species sets limits to its evolution, and thus that internal organizational conditions may (sometimes or always) partially determine the avenues of evolution. The genetic system is not merely a product of Darwinian selection of haphazard mutations in particular environments; its form has in certain respects been determined by the prior necessity of internal coordination. Thus around

82

1950 the conception of a new mechanism of selection had already become rather clear to four thinkers independently. It is striking evidence of the slowness of advance of radically new ideas that fifteen years have had to pass from my first formulation in 1949 to the more extensive discussion of this book.

(N.B. I did not refer to my 1949 formulation in my *Science* (1960) and *Acta Biotheoretica* (1964) discussions because, in the interests of objectivity, I did not wish to prejudice the consideration of so far-reaching an idea by seeming to claim any personal priority or by associating it with other speculative ideas contained in my 1949 book.)

The following two quotations are instructive as the first author recognized the possibility of internal selection but considered it doubtful, whereas the second left it open as of possible importance.

C. H. Waddington (1957). "There has been at least one suggestion that processes of selection may sometimes occur at the level of the gene itself . . . these cases could perhaps be considered to provide examples of a category of selection which operates, not on the phenotypic results of the developmental activities of genes, but directly on the genotype itself. . . . These examples are, however, not very convincing."

J. B. S. Haldane (1958). Following Spurway, he considered a "selection based on genotypes," a "directional evolution" not due to environmental changes, and the possibility that "the disturbance of a particular developmental process is more or less harmless in one species, but lethal or sub-lethal in another closely related one."

An examination of these passages is instructive. Smuts reached his notion from a holistic philosophy of organism, but left it vague and non-structural. Dobzhansky may have been the first geneticist explicitly to reintroduce the conception of a possible control by the genetic system of its own directions of mutation, after Morgan's repudiation of this idea in 1919. Stern observed what he interpreted as an internal selection, in this case leading to contrasted genotypes with similar phenotypes. Spurway inferred from observations of homologous variation the likelihood of a restricted "mutation spectrum." I recognized the possibility of an internal selection, expressing an internal tendency towards stability, and playing an important guiding role in phylogeny. Bertalanffy, an organicist philosopher, maintained the operation of a triple internal restriction on phylogenetic changes: by the permissible variations in genes, by the genetic system during development, and by the general laws of organization. Lima-de-Faria inferred the presence of internal selection acting directly on the chromosomes from the study of chromosomal gradients and field effects, but perhaps went too far in regarding it as the "prime" determinant of evolutionary trends. Waddington, drawn to the problem by concern with the relations of development and evolution, considered the possibility of one form of internal selection, but regarded the evidence as inadequate. Haldane treated internal selection as a potentially important issue.

Both Dobzhansky and Haldane reached the idea through consideration of the observed differences in mutation rates, these being, it seemed, genetically determined in certain cases. From this it is a small step to the partial genetic determination of the successful directions of genotypic change. Spurway's remarks are noteworthy, since as early as 1949 she gave clear expression to a hypothesis regarding the general importance of internal factors, reached from an analysis of examples of homologous variation, which she has since developed. Soon after, Lima-de-Faria ascribed an even greater role to internal factors.

It is impossible without more extensive discussion than would be appropriate here to be just to all biologists who have discussed internal selection or closely related ideas. For example in several passages of his *Evolution of Genetic Systems* (1939) C. D. Darlington emphasized the presence of a selection acting "on the genetic system at every level, gene or chromosome, cell and individual" and spoke of a "gene struggle as well as a cell struggle." "The capacities of a genetic system for evolutionary change are limited by its properties at every level. . . ." Yet he did not suggest that the evolution of genetic systems involved a new non-Darwinian mechanism of selection.

Internal Factors in Evolution

J. Langridge (1958) proposed a hypothesis of developmental selection in which internal factors, i.e. barriers imposed by ontogenetic development, determine the survival of mutants.

J. H. Woodger (1959) stressed the importance of random union and *random development* in obtaining the Mendelian ratios, a crucial point because all genotypes do not enjoy equal probability of successful development. This is obvious and Mendel was aware of it, but it is valuable to have it stressed in a formal treatment.

In 1960, encouraged by J. B. S. Haldane's observations (1958), I reformulated the principle.

As examples of recent attitudes the following may be quoted:

A. C. R. Dean and *C. Hinshelwood* (1963). ". . . the selection of an appropriate reaction pattern by self-replicating matter may be just as important as the selection of special individuals."

C. Loring Brace (1963). Recent work provides "the basis for the possibility that the process whereby variation is produced may also determine the direction of the variation when (external) selective factors are inoperative."

These quotations represent, in this respect, the most advanced thought of the period 1940 to date. Many other discussions of the elimination of deleterious mutations during development implicitly involved the consequence that internal factors restrict the successful directions of evolutionary change and came to the threshold of this inference, but without passing over and making it explicit, or considering its possible general importance. For example, Schmalhausen (1949) based his theory of stabilizing selection on the ability of the mechanism of individual development to undergo changes independently of the adaptive properties of the phenotype, but did not infer that this implied restrictions on the successful changes leading to new evolutionary steps. Lerner (1954) considered developmental or ontogenetic homeostasis, but his concept of genetic homeostasis is a population property. Waddington has emphasized the restrictions imposed by developmental conditions, but without inferring that this implied a new non-Darwinian mechanism of selection. Moreover, the influence of internal factors was not mentioned as being of possible evolutionary importance in any of the leading Darwin Centenary Surveys, published 1959–60, though it had been the subject of private discussion during the 1950's, e.g. by readers of Spurway, Bertalanffy, and Lima-de-Faria.

Thus while many *special* conceptions had been developed (developmental elimination of deleterious mutations, factors influencing mutagenesis, differential mutation rates, mutator genes, developmental homeostasis, developmental channels or barriers affecting the results of mutations, etc.) few attempts had been made to draw the *general* conclusion that one of the basic postulates of the synthetic theory was too restricted: the assumption that the variations undergoing external selection were undirected in relation to phylogeny. For example, if internal selection is effective it is not necessary to ascribe all the biologically significant properties of the genome (e.g. linkage of genes with related functions) to past adaptive selection. They may be a direct consequence of the C.C. and of internal selection.

It will be seen from this survey that many biologists, approaching the theme from several points of view, began seriously to consider the role of internal factors from around 1949. But no published analysis came fully to grips with the fact that a new class of directive factors in evolution had emerged, that this implied that the synthetic theory only covered one aspect of the historical facts, and that such a far-reaching conception called for a searching re-examination of the position both by biological philosophers and by the many different kinds of specialists concerned with genetics and evolution.

6: Difficulties Answered

The present situation is marked by a paradox of a kind frequent in the history of science: the operation of internal factors in several special contexts is already regarded as a commonplace, particularly in private discussions, but the fact that this contradicts an assumption of the synthetic theory of evolution is neglected in the literature, this being for many an emotionally charged issue.

The elimination of deleterious mutations during development was recognized during the 1930's. Yet the influence of internal factors in determining what can constitute successful mutations (and so partially determining phylogeny) is only in the 1960's becoming widely recognized as an important issue. This thirty-year lag is instructive. The main reasons may have been: concentration on observations rather than theory; a continuing stress on statistical and ecological aspects with consequent relative neglect of the internal structural developmental processes; the absence of direct evidence regarding the character of successful mutations leading to new differentiations; and the difficulty—until the conception of the C.C. had been formulated—of achieving a valid theoretical and observational separation of internal from external selection. These influences combined to stifle discussion of the relation of internal selection to phylogeny.

Yet these factors alone do not account for the remarkable fact that the Darwinian surveys of 1959 neglected to mention that some eight years earlier four writers had formulated ideas which implied that the Darwinian type selection might have to be supplemented by another. In an endeavor to find the explanation I have had discussions and correspondence (1959–1963) with a number of leading evolutionary biologists and members of the new school of molecular biology. Their responses to my arguments may be summarized in the following representative statements:

"This is a complete misunderstanding; there is no internal selection."

"Modern evolutionary theory rests entirely on the statistics of populations."

"If there is any internal selection, there is no basic distinction between it and Darwinian selection."

"The idea of internal selection may be valid, but it is of no immediate practical value; it is premature; we know too little."

"The idea is a commonplace; it is obvious to anyone aware that organisms are organized. It is not of any special interest."

"There is nothing new in the idea of internal selection; I have been teaching it for years; it should be in the books by now."

"To a molecular biologist the idea is obvious. But why bother? The geneticists and evolutionary theorists must come round soon. The facts will speak for themselves."

"As a molecular biologist I consider the argument obviously correct, important, and timely. For some reason, perhaps because we know so little about evolutionary theory, we molecular biologists are over-cautious in drawing evolutionary consequences from our ideas."

Such variety of opinions is healthy for science; a conflict of views is necessary to provoke the difficult reconsideration of fundamental assumptions. What is instructive is the slow percolation of new ideas across the street from one department to another, from molecular to evolutionary biology.

My study of opinion on this issue has shown that the deepest reason for the failure of geneticists and evolutionary theorists to recognize and discuss internal selection in print as a new non-Darwinian mechanism is that they have not yet been compelled to do so. *None of them yet recognizes the importance of internal selection as a fundamentally novel principle potentially affecting the entire interpretation of the history of life on this planet.* No dramatic event has yet focused attention on the idea and forced them to go through the uncomfortable process of questioning long-accepted assumptions and of tracing the consequences of a new idea. For every ten who vaguely feel the significance of the idea, there may be only one who has a vivid sense of its far-reaching importance and also the time and energy to do something about it. Hence this lay attempt to provoke specialist discussion.

Perhaps the most instructive response that I have received is the suggestion that the idea is "commonplace," indeed a "truism." But a "commonplace" is a trite or trivial matter of no interest, and a "truism" a thesis which no one disputes. Yet *all textbooks still assert that Darwinian selection is the directive factor in evolution,* and only biologists who fail to appreciate the revolutionary implications of the idea and the new fields which it opens to research can suggest that the operation of internal factors has no interest for them.

But it is not merely inertia that delays full acceptance of the idea. Even those who have followed the argument thus far may none the less find in their minds two recurrent obstacles to accepting its novelty and importance:

1. *"Since there is universal interaction, the two selective processes are not separable. Internal and external environment, ontogeny and phylogeny, are aspects of a single total story: the history of the changing forms of life guided by one comprehensive process of selection. A distinction between external and internal selection has no fundamental validity."*

92

Difficulties Answered

Those who have read thus far and can still think like this have not yet appreciated the revolutionary importance of the conception of *structural ordering*, which requires a kind of thinking entirely different from that appropriate to ecology. Organisms are finely structured and coordinated units, and there is a basic theoretical distinction between the inside of an organism and its external environment. Moreover, as we have seen, external selection is comparative, statistical, and competitive; internal selection is intrinsic, singular, and coordinative. No distinction could lie deeper. Confusion can only arise from treating together the competition of free individuals and the harmonious cooperation of the ordered parts of a unit. The clarity of biological thought demands attention to this contrast.

In terms of mathematics, the kernel of external selection is *differential reproductive efficiency;* that of internal selection is *satisfaction of the C.C.*, and this will soon be capable of being directly observed.

To neglect this distinction any longer would hinder the advance of biology at the very point where progress towards structural and quantitative precision is most possible at the present time. Whereas no increase of knowledge which we can now anticipate is likely to throw much light on the exact meaning of fitness for some hypothetical ecological niche and its associated competition many million years ago, structural studies of the coordination of living cells should lead to a

93

deepened understanding of the C.C. within, say, a generation. When the C.C. are known (see Chapter 8) and the structure of particular successful mutations can be specified, the consequences of internal selection will be open to scientific prediction, experimental control, and direct observation.

Some may find more force in a second objection:

2. *"The idea is premature and the evidence for it too small. It is of no use to working biologists today."*

It is true that at this moment (1964) too little is known of the organization of the cell nucleus as a whole for one to be able to picture in detail particular deviations from its normally coordinated state. Those mutations which have been most studied correspond to changes affecting the production of particular enzymes rather than the mode of coordination of the nucleus or cell as a whole. Since no favourable mutation affecting the general coordination has yet been identified, it is not possible to give specific examples of the operation of internal selection.

But in the history of science there are periods when the logic of a new idea must be developed *before* the corresponding experiments and observations can be undertaken. At such a time those who can seize the emerging idea and strengthen it by applying it in their work may achieve important advances.

There are also several realms in which the idea

of internal selection is already proving of value to working biologists:

1. There is considerable evidence that the genetic system determines the relative rates of certain kinds of mutations. If so, it is probable that the genetic system also determines the main directions of possible change. Indeed this is now widely assumed.

2. Since 1940 many workers (e.g. Stern, 1943; Spurway, 1949, 1960; Lima-de-Faria, 1952, onwards; Langridge, 1958; Sondhi, 1961; and others less explicitly) have found themselves led to interpret particular genetical observations in terms of the influence of internal factors (such as premutational conditions, genetic control of mutations, mutator genes, genetic and developmental homeostasis, developmental barriers, etc.) restricting the possible lines of evolutionary change. For example, Lima-de-Faria (1962) has been led by cytological observations to stress the importance of the criterion of "compatibility with the system of order which is the chromosome field."

Though it may be difficult to anticipate how a particular perturbation or transformation of the developmental processes will finally be traced to a specific locus, or to chromosomal or cellular structure, many techniques are already converging on this task, which is crucial to the theory of ontogeny. Within a generation it may be possible to produce selected mutations of known genetic structures in particular species and to observe the elimination of some, and the successful operation of others, during the processes of development.

3. Another line of research which is leading towards the identification of the C.C. as they affect the genotype and its mutations is the statistical study of non-random arrangements in biological macromolecules. Recent biochemical and theoretical studies (e.g. R. V. Eck, 1961, 1962; H. H. Pattee, 1961) point to restrictions on the possible sequences in the linear biomolecules expressing specificity. Pattee considers that there is an important element of order in such macromolecules which is a precondition, not a consequence, of adaptive evolution. Certain changes in the order in biomolecules may be more stable or probable than others. Such constraints form part of the C.C. and studies of coding must ultimately throw light on the C.C. Coding determines ordered specificities which are not arbitrary, but must satisfy the C.C. This fact may have important implications for the application of information theory to organisms.

When the C.C. have been partly or wholly identified and when knowledge of the structure of chromosomes and their immediate environment is sufficiently advanced, it may even become "possible to predict the class of mutations which is capable of surviving internal selection in a given species in a constant environment" (Whyte, 1960).

This evidence suggests that further advances in structural knowledge requiring interpretation in terms of internal selection and of the C.C. may be expected soon. If this is so, we may again recall what Einstein stressed: that given a sufficiently powerful formal assumption a fertile and comprehensive theory may sometimes be constructed without prior attention to the detailed facts or even before they are known. If biology is ever to possess a truly comprehensive theory, it must, like those of physics, rest on comprehensive formal assumptions such as this: Organisms are necessarily so highly co-ordinated that only a restricted (and ultimately definable) set of variations from any starting point are permissible. The structure of organisms determines the main avenues of phylogeny. This may be one of the postulates of the lacking theory of variations.

Even if all these arguments are acceptable, there may remain a final reluctance to take them seriously. Is this emphasis on internal factors leading biology towards a holistic or finalistic philosophy of organism *of a kind which can prejudice its objectivity and precision?* This is not so, because a theory of internal factors, complementary to external ones, invites exact analysis and test, and does not limit the causes of evolution to any one class of factors.

The operation of internal factors as here conceived is not *vitalistic,* since it involves potentially observable structural parameters; nor *orthogenetic,* since it allows many avenues of potential evolution; nor *nomogenetic*

Internal Factors in Evolution

(Berg, 1926) since it is complementary and not alternative to the operation of external factors. Attention to internal factors is a natural consequence of the organicist view which regards the characteristics of organisms as consequences of a state of organization in complex structured systems of certain kinds in appropriate environments. On this view the genotype may not only be a self-regulating and self-repairing system but in some degree determine its own viable transformations.

7: Conclusions

The argument has led to the following conclusions which are believed to be reliable in the light of existing knowledge, though their formulation is likely to prove imperfect:

1. There is no justification for the common assertion that a selective process of Darwinian type is the only directive agent in phylogeny. On the contrary there are good grounds for inferring that there also operates an internal selective process acting on genotypes and their early consequences in terms of their inner coordination.

2. Internal selection does not act by causing a differential survival of two competitive types, but passes or rejects each mutated genotype by itself. The St. Peter of internal selection does not make invidious comparisons, but judges each candidate on its own merits and reforms some.

3. Internal selection may prove to have been most important, and perhaps solely operative, during some or all of the major steps in evolution, which are not yet understood. The *primary* cause of the emergence of a higher category may not always have been the move-

ment of a population into a new adaptive zone, but sometimes the occurrence of an organizational transformation of the genotype resulting in a phenotype capable of exploring a new zone.

4. Internal selection, as here defined, may in special situations go beyond the passive selection of given variations, and by the reformation of inappropriate disturbed genotypes exert a direct molding influence guiding evolutionary change along certain avenues.

5. Many biologically significant properties of the genetic system (linkage of genes with related properties, co-adapted harmony of the genetic and developmental system, etc.) may be the result, not of past external selection, but of internal selection.

6. The study of phylogeny in terms of population statistics must be complemented by an analysis based on the structural ontogeny of the individual. Structural principles may be as powerful in directing historical change as the statistics of groups, and a typology of structural organization may be indispensable to the understanding of the forms of life. The puzzling stability and discontinuity of species phenotypes may be the consequence of the existence of a discrete spectrum of solutions to the C.C. More specifically, the hierarchical organization of a discrete manifold of organic types may be explained as solutions to the C.C.

7. Other problems which are still obscure in the synthetic theory, e.g. production of non-adaptive char-

acters, may be clarified by a theory of internal factors allowing these problems to be examined afresh from a new point of view.

8. One of the main benefits of studying internal selection is that thereby more can be learned about biological organization, i.e. about the C.C. Until the C.C. have been identified, no theory of phylogenesis, of ontogenesis, or of their relations, can be regarded as definitive. Moreover, the C.C. hold the clue to the relation of physical laws to organic processes and to the unity of the organism.

9. Just as domesticated species and artificial selection gave Darwin indispensable clues for his theory, so controlled mutations alone can lead to a definitive theory of internal selection.

10. Internal selection is a phenomenon appropriate both to the intellectual situation and to the experimental techniques of the late xxth century. Smuts dimly anticipated internal selection around 1925; it was formulated as a scientific conjecture by four workers around 1950; it is likely that it will be accepted as an independent mechanism of evolution by the majority of competent specialists by 1975. These dates are meaningful; they express the attention paid by this century to organized structure. In the 1960's we are entering the period when attention to internal selection can be especially fruitful.

11. The purpose of a general exposition such as the

present essay is to encourage specialists to consider the application of the idea in their respective fields and to invite future expositors of the synthetic theory to avoid making exaggerated claims for the Darwinian selective process.

12. The timeliness and immediate value of this analysis are evidenced by the questions to which it leads. Some of these are discussed in the next Chapter.

8: Questions for Tomorrow

To one who has watched the advance of fundamental physics and biology over the last fifty years it appears probable that before the end of the century exact science will have entered a new phase through the discovery of the appropriate methods for representing complex ordered systems in course of change. This may bear fruit in two achievements: a unified theory of physical particles and an elegant theory of biological organization. It may well be that *inequalities* and *one-way processes,* being logically more powerful, may also prove theoretically primary and that the equations, reversible processes, and stationary states of current theory will be seen to apply only under restricted, limiting conditions.

These are questions for the generalized physics of the future, but it is within such a vista of advancing physical ideas that we must view the problems of biological organization, the C.C., and their bearing on the theory of evolution. We cannot today anticipate the precise form of the C.C., but we can analyze their task and take a step towards seeing how it may be accomplished.

It will be recalled that the C.C. are the algebraic expression of the 3D geometrical and kinetic conditions determining the spatial and temporal relations of parts and processes in all viable organisms. Our assumption here is that the conception of the C.C. and their actual form will one day be as commonplace to structural biologists as Maxwell's equations are now to physicists. The C.C. are out there in nature, just as much as Maxwell's equations. What can be said about them today?

They are certainly set in 3D space, not in some higher abstract manifold. They are scale-fixed, i.e. contain natural units of length and time (or the equivalent), though in many circumstances organic structures and processes vary greatly in spatial and temporal scale. Moreover, the C.C. probably employ collective parameters, i.e. global variables associated with organic systems as a whole and their internal relations, and not with ultimate parts. The C.C. must be capable of representing both relatively stable structures and the formative processes which generate them.

These features may be summarized in the following requirement: the C.C. must represent organic systems as units which sustain and extend the unified ordering of specific heterogeneity. The simplest solution, or most general realization of the C.C., represents the simplest bio-unit, a hypothetical or actual simplest unit of life, containing within itself in least differentiated form the minimum essential for an organism—perhaps a proto-structure representing the nucleus, cytoplasm, and wall of the earliest cells.

The C.C. define the primary features of biological organization. This covers a wide range of properties, such as:

> The *characteristic roles in relation to the C.C.* of (i) particular chemical atoms, (ii) particular ions and molecules, (iii) particular forms of DNA, RNA, and proteins, (iv) the nucleus cytoplasm and various organelles of the cell, i.e. the functional significance of these entities in the unified activities of the cell and organism. This implies a universal morpho-biochemistry.
>
> The *unified spatial ordering* of differentiated structures, and the *unified timing* of correlated processes.
>
> The one-way *developmental* (replication, growth, life-cycle) and secondary *stationary* (cyclic, homeostatic) aspects of the unified processes.
>
> The dominance of *morphogenetic or ordering* (potential energy) processes over *disordering* (entropy) processes.
>
> The *phylogenetic avenues* determined by mutations conforming to the unified system.

This covers a vast field of problems which are of great complexity when analyzed in the traditional manner. But there are grounds for expecting that they can be resolved by an elegant method, perhaps by one set of C.C.

For all these problems concern *aspects of the unity of organism* and their confusing complexity reflects our failure to identify the precise character of that unity. If organisms are highly ordered and are all variants of a single underlying type—and what scientist who shares with his greatest predecessors the passion for unity can doubt this, or question the value of first assuming it?— then once this order is identified the pervasive logic of organism will guide our minds, the natural avenues of organic structure will open before us, and the actual complexities of organisms, such as the immense numbers of contrasted parts, will no longer intimidate the theoretical intellect. Once in possession of the true bio-logic everything in organisms will seem natural, inevitable, and relatively simple.

There is evidence that this intellectual emancipation lies not too far ahead for conjectures to be useful now. Thus molecular biology is in sight of a universal biochemistry or basic unity of metabolic organization and of its historical evolution; the nuclear genetic code is nearly solved; the search has begun for a basic excitatory and functional process combining energetic, electrical, chemical, and mechanical aspects; this implies that the various differentiated functions may be components of a primary process in which all these are unified; the efficient coordination of distant effects may be achieved by differentiated channels permitting extended electrical actions. These are not fanciful conjectures, but legitimate, and indeed urgent, theoretical aims, once the unity of life as a pervasive type of ordering is postulated.

The signs point towards a comprehensive biological synthesis on a structural basis, complementing the synthetic theory of evolution, perhaps within this century. This synthesis must unify structure and process.

If that be so, what is missing? Individual structural facts covering *stationary* aspects are being collected at an unprecedented rate, but the basic theoretical principles of a structural and *dynamic* synthesis are lacking. What character must these principles possess? What must be their aim?

The answer is surely beyond doubt. What is lacking is *a mathematical conception of the character of the unity of organism, and of organic subunits,* a conception of the mathematical form of the type of unified ordering of processes which is life. Until that mathematical form is identified, the "unity of organism" remains a vague, intuitive, heuristically infertile, macroscopic conception. But once the form of the C.C. is known, a fresh chapter in the history of ideas will be opened and biology will be guided by a new paradigm. It is the C.C. that are missing. They must now be the object of deliberate convergent attack.

Traditional analytical, atomic, and information-theoretical methods tend to neglect the asymmetrical relations which are indispensable in representing such biologically fundamental properties as the *dominance* of more over less stable structures, the spatial *asymmetries* which initiate processes, and the *one-way processes* which generate forms. What is needed is a finalistic principle of the global one-way ordering processes in organisms, the tendency towards the development of form expressed as a collective property of partly ordered parts. Once this principle is given its appropriate expression the stale antithesis of mechanism without tendency and unity without structure will finally be forgotten, its place being taken by organic mechanism possessing an inherent tendency towards order.

The C.C. may be known before there is a base on the moon. But the theoretical foundations of the C. C. will not be the quantum mechanics of 1930, or the biophysics or biochemistry of 1964, but these sciences transformed by generalization into a more powerful theory of complex forms of unified changing order.

In the meantime we can consider more closely the relation of the C.C. to the role of internal factors in evolution.

We must regard the C.C. as *general* conditions which can be satisfied by an immense variety of *particular* forms, just as equations may possess a multitude of particular solutions. In fact the C.C. must possess a great hierarchy of particular solutions corresponding to all the taxa: the phyla, orders, etc. Moreover each particular solution must cover a hierarchy of levels of organization within the organism it represents, and this entire hierarchy must be represented in the zygote, which is the condensed record of the genetic information contained in the adult. In the process of growth this hierarchy, characteristic of the species or individual, is spatially *expanded* by the replication of component structures. The concentrated microhierarchy of the genetic system thus becomes manifest as the macrohierarchy of the adult. This developmental process is theoretically primary; all the special stationary aspects, the homeostasis, cyclic functions, etc., are to be treated as secondary consequences of this basic ontogenetic one-way development.

This interpretation suggests that the genome is itself in some sense hierarchically organized, that there is a meaningful manner in which mutations can be graded in terms of their organizational importance, say from megamutations involving changes in the entire mode of the organization of the cell to the single gene or point mutations which correspond to minor modifications in the DNA coding for the production of a single enzyme. This implies, in turn, as we have already seen, the existence of "selection rules" determining which mutations from an initial genotype are permissible, i.e. can be carried through without undue risk of a fatal disturbance to the entire system.

One purpose of this work is to invite the attention of specialists to what appear in the light of this analysis to be some of the more important questions confronting the theory of internal selection:

  1. Arguments of high generality render it probable

that internal selection has played a significant role in phylogeny. How far is this conclusion supported by direct evidence from particular taxa, e.g. from micro-organisms, plants, and animals? What light does bio-chemical and cytological evidence regarding nuclear and cellular organization throw on this question?

2. When is internal selection likely to be of over-riding importance? Does it play a part in macrogenesis, or, more generally, in the production of favorable mutants and of speciation?

3. Are the major evolutionary advances associated with changes permitted by internal selection in the over-all organization of the chromosomes, cytoplasm, or cell cor-tex, or of the relations between these three groups of factors? Do single genes (cistrons) determine ultimate units of specific structure, and cytoplasmic and cortical factors determine morphogenesis, i.e. 3D orderings of contrasted units?

4. In view of the unique theoretical scope of chiral (skew or screw) forms which are capable of determining both polar-electrical and axial-magnetic correlations, and of the prevalence of helical arrangements and left-handed amino acids in organisms, are the C.C. related to helical pulsations, i.e. to change in helical angles?

5. Do all vital processes involve pulsating structures, undergoing simultaneous cycles of geometrical deforma-tion and electro-magnetic polarization? If so, do mutated genotypes have to be capable of conforming to such pulsations?

6. Are the C.C. the expression of an ordering tendency,

related to the movement towards minimum potential energy of complex low temperature systems, which can account both for the genesis of life and for its maintenance?

7. Is it possible to advance from the coding of stationary and disordered specificity (without a dynamical or coordinating principle) to a more comprehensive coding of morphogenesis and ontogenesis under the C.C.? What light do current theories of coding throw on the C.C. and on internal selection?

The approach to these questions can be aided by a reconsideration of the ancient analogy and contrast between organisms and crystals. For the C.C. are for organisms what the mathematical conditions (3D arrays of equivalent units equivalently placed) determining the 230 crystal groups are for crystals. This parallel helps us to explore the character and role of the C.C. in advance of their discovery:

	CRYSTAL	ORGANISM
Determining Conditions.	Ideal crystals: arrays of identical units equivalently placed and oriented.	C.C. (Unknown)
	Uniform ordering of identical units.	Unified ordering of contrasted parts.
Solutions.	230 discrete solutions, classifiable as a hierarchy of	Myriads of discrete solutions, classifiable as a hierarchy

Internal Factors in Evolution

	CRYSTAL	ORGANISM
	systems, classes, and groups.	of phyla, classes, genera, species, and variations.
Temporal character.	Static.	One-way process, with cyclic aspects.
Mathematical aspects.	Group theory. Symmetry, symmetrical relations, equations, metric.	(Unknown.) Nonequivalence, asymmetrical relations, inequalities, order.
Character of units.	Symmetry type represented in each unit cell.	? The C.C. type represented in the organism and in all functional subunits.
Transformations.	"Mutations" (e.g. under changes of temperature) from one symmetry type to another.	Mutations from one solution of the C.C. to another neighboring solution.

It is not the mere expression of an idea which counts in science, but the strength of the belief that it is important so that action follows; the realization of why it is important and what its implications are; and the understanding of why it is timely, which make its expression capable of leading to discoveries at a particular period in the history of science.

I believe that the principle of internal selection is important and timely because it is the expression of the contemporary experimental and theoretical concern with structure. If that is so, then those that apply their energies to clarifying and applying the principle are likely to be rewarded by discoveries relating to the co-ordinative conditions which characterize living systems.

References

Bachelard, G., *Le Materialisme Rationel* (Paris, 1953), p. 146.

Berg, L. S., *Nomogenesis, or Evolution Determined by Law* (London: Constable, 1926).

Bertalanffy, L. v., *The Problem of Life: An Evolution of Modern Biological Thought* (London: Watts, 1952; New York: Harper's Torch Books, 1960).

Brace, C. L., American Naturalist xcvii (1963), p. 39.

Darlington, C. D., *The Evolution of Genetic Systems* (Cambridge: The University Press, 1939).

Dobzhansky, T., *Genetics and the Origin of Species* (New York: Columbia University Press, 2d Ed., 1941).

Eck, R. V., Nature *191*. 1284. (1961).
            Jour. Theoret. Biol. 2. 139. (1962).

Haldane, J. B. S., Journal of Genetics. *56*. 11. (1958).
            In *Darwin's Biological Work*, P. R. Bell, Ed. (Cambridge: The University Press, 1959).

Dean, A. D. R., and Hinshelwood, C., Nature. *199*. 7-11. (1963).

Huxley, T. H., "Letter to G. J. Romanes" (1888). See A. Keith, *Concerning Man's Origin* (London: Watts, 1927).

Langridge, J., Australian Journal of Biological Sciences. *11*. 58-68. (1958).

Lerner, I. M., *Genetic Homeostasis* (New York: Wiley, 1954).

Lima-de-Faria, A., Chromosoma. *5*. 1. (1952).
　　　　　Chromosoma. *6*. 330. (1954).
　　　　　Hereditas. *42*. 85. (1956).
　　　　　Jour. Theoret. Biol. 2. 7. (1962).

Medawar, P. B., *The Future of Man* (London: Methuen, 1960), p. 38.

Morgan, T. H., *The Physical Basis of Heredity* (Philadelphia: Lippincott, 1919).

Oparin, A. I., *Life, Its Nature, Origin and Development* (Edinburgh: Oliver and Boyle, 1961), p. 82.

Pattee, H. H., Biophysical Journal. *1*. 683. (1961).

Pledge, H. T., *Science since 1500* (London: Science Museum, 1939).

Schmalhausen, I., *Factors in Evolution* (Philadelphia: Balkeston, 1949).

Smuts, J. C., *Holism and Evolution* (London: Macmillan, 1926).

Sondhi, K. C., "Developmental Barriers in a Selection Experiment," Nature. *189*. 249. (1961).

Spurway, H., "Remarks on Vavilov's Law of Homologous Variation" in Supplemento. La Ricerca Scientifica (Pallanza Symposium) *18*. Cons. Naz. delle Ricerche. (Rome: 1949).

Stern, C., "On Wild-type Iso-alleles in Drosophila Melangaster." Proceedings, National Academy of Science. *29*. 361. (1943).

Vandel, A., Anné biologique II, p. 179. (1963).

Waddington, C. H., *Strategy of the Genes* (London: Allen and Unwin, 1957).

Weismann, A., *On Germinal Selection as a Source of Definite Variation*. 2d Ed. Transl. from German Edition, 1896 (London: Religious and Science Library, 1902).

Whyte, L. L., Science. *132*. 945, 1964-5. (1960). Acta Biotheoretica Vol. XVII, 33. (1964).

Woodger, J. H., *Studies in the Foundations of Genetics in the Axiomatic Method*. Edited by Henken, Suppes, and Tarski (North Holland Pub. Co., 1959), p. 423.

Other relevant papers by L. L. W.

"One-way Processes in Biology." Proc. Xth Int. Congr. of Philosophy (Amsterdam, 1948), p. 298.

"Note of the Structural Philosophy of Organism." B. J. Phil. Sci. *5*. p. 332. (1955).

"One-way Processes in Physics and Biophysics." B. J. Phil. Sci. *6*. p. 107. (1955).

"On the Relation of Physical Laws to the Processes of Organisms." B. J. Phil. Sci. 7. p. 347. (1957).

Recent Expositions of the Synthetic Theory

Huxley, J. S., *Evolution, the Modern Synthesis* (London: Allen and Unwin, New Edition, 1963).

Mayr, E., *Animal Species and Evolution* (1963).

Simpson, G. G., *The Major Features of Evolution* (1953).

Index

Index

Index

Index

Index

Index

Populations, 43, 66
  statistical theories of, 64-65, 68, 90, 100
Proteins, 34, 37, 50, 65, 105

Quantum theory, 16, 55, 108

Radiative fluctuations, 52
Randomness, 45-46, 52, 53, 77, 86
Rensch, B., 41
Reproductive differential, 44, 47
Reproductive rate, 43
RNA, 34, 37, 105
Roux, W., 11, 75, 76, 77

Schmalhausen, I., 87
Science, 20, 83
Science:
  history of, 89
  methods of, 9-25 *passim*
Selection, 19, 21, 65
  adaptive, 58-59, 64, 69
  external, 44, 58, 60, 69-70, 78, 92, 93
  internal, 49-73 *passim,* 77, 80, 84, 85, 88, 89-98 *passim,*
    99, 100, 101, 112
  "natural," 48, 58
  passive, 100
  theories of, 42, 90-91; *See also:* Mutation; Variation
Simpson, G. G., 41, 44
Smuts, J. C., 80, 84, 101
Solar system, 17

Index

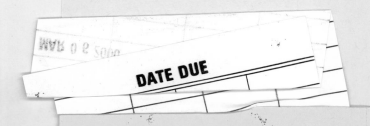